Digital Human Modeling for Vehicle and Workplace Design

Don B. Chaffin

with contributions by
Cynthia Nelson
John D. Ianni
Patrick A.J. Punte, A.J.K Oudenhuijzen, A.J.S. Hin
Darrell Bowman
Deborah Thompson
Brian Peacock, Heather Reed, Robert Fox
D. Glenn Jimmerson

INTERNATIONAL®

Society of Automotive Engineers, Inc.
Warrendale, Pa.

eISBN: 978-0-7680-4887-2

Library of Congress Cataloging-in-Publication Data

Chaffin, Don B.
  Digital human modeling for vehicle and workplace design / Don B.
Chaffin with contributions by Cynthia Nelson ... [et al.].
    p. cm.
  Includes bibliographical references and index.
  ISBN 0-7680-0687-2
  1. Motor vehicles—Design and construction. 2. Human engineering—
Data processing.
  3. Human-machine systems—Computer aided design. 4. Human-machine
systems—Mathematical models. 5. Digital computer simulation. I. Title.

TL250 .C49 2001
620.8'2'0285—dc21   00-069822

SAE Order No. R-276

Contents

Foreword

The ability to digitally simulate how humans interact with a product has the potential to revolutionize the way companies design, build, operate, and maintain new products.

Digital modeling and simulation techniques have already proven their ability to significantly reduce the cycle time and cost of designing new products, and have generally improved the quality of products and made them faster, easier, and cheaper to produce, operate, and maintain.

But many products—such as high performance aircraft and spacecraft—present additional design challenges in human factors. To get the greatest performance, comfort, and safety from these products, engineers need to know early in the design process how effectively and efficiently humans will be able to interact with them.

These challenges becomes increasingly important as businesses expand into global markets, where the success of new products depends on accommodating a greater diversity of human sizes, shapes, and other physical characteristics.

Through accurate digital modeling and simulation of human interactions with a product, all stakeholders in the development of the product—marketing, engineering, production, and support—can visualize the design and share their concerns about it and make suggestions for improvement.

Such evaluations will allow design decisions to be based on a broader understanding of the user before development resources are committed, and will eliminate many additional—and typically unnecessary—steps that often occur later in the development and support processes.

A simple evaluation of manual and visual access for adequately installing and inspecting components in a product, for example, can save many unnecessary steps and thousands of labor hours later on, during the production and maintenance of the product over its lifetime.

So in an age focused on getting better, faster, and cheaper results through computer technology, the development of effective digital human modeling capabilities seems to be a logical next step for dramatically improving the cycle time, quality, and cost of producing ergonomically efficient designs.

For those companies that value ergonomic thinking and taking the perspective of the customer, assembler, maintainer, and operator during the design process, investments in hardware, software and training for digital human modeling and simulation will pay big dividends.

We must keep in mind that, while the future may belong to those who plan, it also belongs to those who implement processes, such as digital human modeling, that improve the speed, quality, and effectiveness of how we think and what we produce for mankind.

In this book, Don Chaffin explores various examples of how digital human modeling has been implemented in different fields. Because these studies show both the promise and the problems associated with this emerging technology, they provide a good foundation for further development in this important field. Accordingly, they should be evaluated carefully and applied fruitfully to integrating this new technology into the design process.

David O. Swain
Boeing Senior Vice President of Engineering & Technology
President of the Boeing Phantom Works

Preface

This book is dedicated to the proposition that one of the most cost-effective means of improving the ergonomic aspect of any future vehicle or workplace design is to utilize the rapidly emerging technology referred to as digital human modeling, or DHM. The first chapter in the book describes the historical basis and development of the most popular DHM programs over the last 30 years. It is shown that some of the early programs were limited in scope to address specialized types of design problems, such as whether a small pilot when operating a new aircraft could reach or see a display or control surface in a proposed design. Other programs were meant to solve a much broader array of possible human-machine interactions, including whether many people of different size, age, and gender could fit, reach, grasp, balance, move, lift, and walk around in virtual environments without undue discomfort or harm. In addition the developers of these more robust DHM programs included task performance time predictors in their software. In other words, the developers of these latter DHM programs were expecting that their software would replace the need for having a well-trained ergonomics consultant when advice was needed regarding how certain groups of end users of a proposed design would be affected by the design. Of course, some of these new software "expert systems" did not come cheap, with prices running as high as $60,000 or more per seat for a full-feature DHM. But given the fact that the world's educational programs are not producing very many well-trained ergonomists (only about 771 individuals are currently fully certified for professional practice in the U.S.), these software programs offer a tempting solution to complex human-machine problems, especially for those companies that already are using various math based design evaluation methods.

After the description of existing DHM programs in Chapter 1, seven case studies are presented wherein digital human models were used to solve different types of physical problems associated with proposed human-machine interaction tasks. The authors of each of these case studies were highly involved in each problem scenario, allowing the reader to gain an intimate appreciation of what each user/author expected to gain from the application of a particular DHM, and what benefits and limitations resulted from their efforts. The overall impression from these cases is very positive in general, but the users/authors admit to having high expectations for the DHM they chose, but in some of the cases they were not able to achieve all of their goals.

The final chapter attempts to summarize the collective experiences and lessons learned from the seven cases. It is my impression that to resolve certain types of questions, such as whether a person of certain anthropometric features can fit into a vehicle, or reach to an object, the existing DHM packages are very helpful. What appears to be more challenging to these users, however, and caused them to take a great deal of their time and effort, was the need to: (1) assemble the necessary databases in the correct format needed to create the required virtual environment for a good DHM simulation, and (2) easily move or position the digitally rendered person in ways that are representative of the population being simulated. It is estimated in a couple of these cases that very significant time and cost savings resulted from using a DHM early in the product design cycle. The cases also illustrate, however, that for more comprehensive ergonomic assessments the user must have a good understanding of ergonomic principles and concepts to avoid misusing or misinterpreting the DHM simulation results.

In general, I believe the case studies are very good examples of how companies can use DHM programs in their attempts to create much more people friendly products and workplaces. It should be clear, however, that the DHM developers must continue to refine their programs to provide users with more flexible I/O data formats. It also is clear that valid DHM predictions of population strengths, postures, motions, discomfort and endurance are not easily acquired and used to guide a new design. It is hoped that the recognition of these deficiencies, as described in these cases, will inspire more research and development to assure that a comprehensive array of valid human attributes will be modeled in the future to assist those wishing to use future DHM programs as part of their CAE and product design efforts.

Don B. Chaffin, PhD, PE, CPE
The University of Michigan
Ann Arbor, Michigan

Acknowledgments

First and foremost, I would like to thank the chapter authors: Cynthia Nelson of the Boeing Company; John Ianni of Wright-Patterson AFB; Patrick Punte, Aernout Oudenhuijzen, and Andrea Hin of TNO (Netherlands); Darrell Bowman of International Truck and Engine Corporation; Deborah Thompson formerly of the DaimlerChrysler Corporation; Brian Peacock, Heather Reed, and Robert Fox of General Motors Corporation; and Glenn Jimmerson of Ford Motor Company. These individuals willingly volunteered their time to produce the seven well-written and nicely illustrated case studies. Without their extra efforts this book would not have been possible.

Of course for any writing project such as this there is always a group of people who are essential to producing a document that clearly expresses the intent of the authors. At the University of Michigan, Patricia Terrell collected the manuscripts and expertly organized the figures and tables.

I also would like to acknowledge Michael Biferno, Manager of Human Factors at the Boeing Company Phantom Works. Mike's outstanding leadership of the SAE G13 Committee on Human Modeling Technology and Standards led to the conceptualization of this book. During the meetings of this committee a project was developed to write the book. Without the G13 organization it would have been very difficult to assemble the people involved and the topics presented in the book.

Lastly I would like to thank my colleagues at the Center for Ergonomics at the University of Michigan for supporting my time on this project over the last two years.

Don Chaffin
July 2000

Acronyms

3DSSPP 3D	Static Strength Prediction Program, by the University of Michigan.
AEF	Air Expeditionary Force.
AFB	Air Force Base.
AFRL	Air Force Research Laboratory.
AMRL	USAF Aerospace Medical Research Laboratory.
ANTHROPOS	Human simulation model from the German firm IST; successor of ANYBODY.
ANYBODY	Human simulation model from the German firm IST; predecessor of ANTHROPOS.
ATOMOS II	An EU sponsored project to enhance safety of operation of ships. Chemical tanker designed for "fast, safe, and economical transportation."
BHMS	Boeing Human Modeling System.
BMD-HMS	Boeing McDonnell Douglas-Human Modeling System.
CAD	Computer aided design.
CAE	Computer aided engineering.
CAEA	Computer-assisted ergonomic analysis.
CAR	Crew Assessment of Reach program.
CATIA	Computer-Aided Three-Dimensional Interactive Application; a CAD system by Dassault Systems.
COM	Center of mass.
COMBIMAN	Model from the Wright-Patterson Air Force Base Human Performance Laboratory.
DENEB	A company specializing in robotics simulation (now part of Delmia Corporation).

DEPTH	Design Evaluation for Personnel, Training, and Human Factors.
DHM	Digital human modeling.
DMA	Digital modeling assembly.
DMU	Digital mockup.
DOT	Department of Transportation.
dVISE	VE software.
EAI	Engineering Animation, Inc. (now Unigraphics Solutions Inc.)
ECDIS	Electronic charts in ship systems?
ERGOMAN	Human simulation model from the Laboratorie d'Anthroplogie Appliguee et d'Ecole Hamaine in Paris.
ERP	Eye reference point.
EVA	Extravehicular activity.
FMCSR	Federal Motor Carrier Safety Regulation
FMVSS	Federal Motor Vehicle Safety Standard
FMVSS 101	Federal Motor Vehicle Safety Standard 101, Controls and Displays.
HMD	Head mounted display.
HMS	Human modeling system.
ICEM-DDN	(Ch. 4)
ISS	International Space Station.
IST	German firm that markets ANTHROPOS.
L4/L5	The intervertebral disk between the 4th lumbar (L4) and 5th lumbar (L5) vertebrae of the spinal column.
L5/S1	The intervertebral disk between the 5th lumbar (L5) and 1st sacral (S1) vertebrae of the spinal column.
LSA	Logistics Support Analysis.
MDHMS	McDonnell Douglas Human Modeling System.
NADC	U.S. Naval Air Development Center
NASA	National Aeronautics and Space Administration.
NBP	Neutral body posture.
NGMT	Next Generation Munitions Trailer.

NIOSH	U.S. National Institute of Occupational Safety and Health.
ORU	On-orbit replaceable unit.
RAMSIS	German model; Realistic Anthropological Mathematical System for Interior Comfort Simulation.
RULA	Rapid Upper Limb Analysis.
SAE G13	Society of Automotive Engineers committee (Human Modeling Technology and Standards Committee).
SAFEWORK	Human simulation model from Genicom Consultants, Inc., of Montreal, Canada (now part of Delmia Corporation).
SAMMIE	A digital human model originally developed by Maurice Bonney and colleagues at Nottingham University.
SMS	Ship movement simulator.
SSP 50005B	A document that is an extraction of the flight crew integration design requirements contained in NASA-STD-3000 Volume IV Revision A tailored to the International Space Station Program (SSP).
STL	Stereolithography.
TECMATH	Software company that participated in the development of the German model RAMSIS.
TEMPUS	Model for simulating astronauts assembling the space station; a result of NASA supported effort within the Department of Computer and Information Science at the University of Pennsylvania in the mid-1980s.
TNO	Abbreviation for Nederlandse Organisatie voor toegepast-natuurwetenschappelijk onderzoek, official Dutch name for the Netherlands Organization for Applied Scientific Research.
VE	Virtual environment.
VR	Virtual reality.
VRML	Virtual reality modeling language.
WATBACK	Model from the University of Waterloo.
WERNER	Human simulation model from the Institute fur Arbeitsphysiologie and der Universitat Dortmund.

Chapter 1

Introduction

Don B. Chaffin, PhD
The University of Michigan
Ann Arbor, Michigan

1.1 Objective of Human Simulation in Design

Fast, high quality computer graphics now allow us to render very lifelike images of people performing a multitude of tasks within various computer aided design (CAD) programs. Furthermore, our statistical descriptions of various population attributes, such as the size, shape, strength, and range of motion of a specific group, have become quite refined. Thus it now is possible to position and move computer generated hominoids, or avatars, as some like to call them, to predict the performance capabilities of designated groups of people within a computer rendered environment. The ease of doing this and the accuracy of the resulting simulations, however, continues to be of concern to those who may be considering the purchase of one of the commercial software packages that provide such human simulations.

The objective of this book is to explore, through seven case studies, how this rapidly developing human simulation technology is being used in practice, and whether it is meeting the needs and expectations of different user groups. In some of the cases it will be very clear that the technology is not well enough developed to meet all of the stated design goals and expectations of the users. We hope that such results will inspire those amongst us who are more research oriented to realize an opportunity to develop new knowledge to enhance future human models. In other cases, however, it will be shown that the technology has been very useful in solving some specific problems during the design of new systems wherein physical human-hardware interface issues were critical.

1.2 Why the Interest in Human Simulations in Design?

A multifaceted reason underlies the increasing importance of human simulation software in the design of hardware systems. In some of the cases it would appear that the organization wishing to employ this technology simply had a top-down commitment in a particular design process to use digital mockup (DMU) methods during the early phase of design. In such cases it was believed by the designers and managers that using human simulation technology within the DMU would decrease the design time and enhance the number and quality of design options that could be rapidly evaluated by the design team. This view is consistent with the concept of reducing the total design and engineering costs by using more computer aided engineering (CAE) and DMU methods to achieve rapid prototype development and testing, as diagrammed in Fig. 1-1.

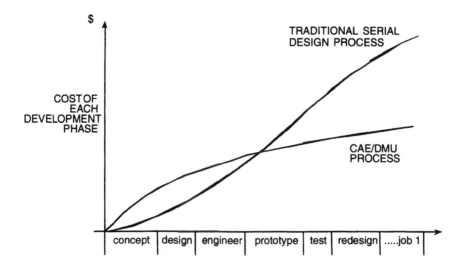

Fig. 1-1 Example comparison of accrued costs associated with either traditional design process (with several physical prototypes being necessary for sequential testing) or use of contemporary computer aided engineering with digital mockup evaluations.

In other cases it would appear that the designers were very concerned that certain ergonomic features had to be correct to accommodate large expected variations in people's size, shape, or strength when operating or maintaining the system being designed. In other words, these designers had acquired enough knowledge about ergonomics to believe that if ergonomic features were not considered in the design, it would adversely affect the safe and effective operation of the system and/or the end user acceptance of a proposed product.

Lastly, it seems that there is a contemporary fascination with the idea of rendering lifelike images within various designs. Perhaps this is an attempt to show that any new hardware design is relevant and acceptable to all people, and that what we do as designers and engineers has a higher purpose than simply decreasing product and operating costs and increasing market share. Perhaps, in essence, we are interested in showing our colleagues and customers how we can improve human conditions in general by inserting lifelike human simulations in our CAD renderings.

1.3 How Do the Roles of Human Simulation and Virtual Reality Technologies Relate in the Design of Systems?

Simulating a person functioning in a particular CAD environment, such as walking, reaching, carrying, or exerting force on an object is quite different than creating 3D virtual images of objects, which then can be perceived by the designer and others to assess whether the object meets certain visual design goals. In the former situation the intent is to provide the designer with the means to determine if people of different age, gender, size, and strength characteristics can safely and effectively perform those essential tasks necessary to operate or maintain the system being designed. In the latter situation, virtual reality or VR, the intent is to provide the designer with a 3D full-sized virtual prototype of a new design so that the designer and others can interact with it. In this context, the digital human modeling and simulation discussed in this book should be seen as a precursor to virtual prototyping. It allows the designer to ask a number of "what if" questions in the early design stage about various people who might be affected by certain features that are considered important in a particular design. Once a design is deemed to meet these simulated ergonomic requirements, then a virtual prototype can be created and evaluated by having groups of potential end users interact with and

3

evaluate the virtual design. As such, both DHM and virtual reality should be considered complementary and necessary technologies today as we strive to use concurrent engineering to achieve better ergonomics in the rapid design and testing of both new products and workplaces.

1.4 A Brief History of Some Popular Human Simulation Methods for Design

Some excellent descriptions of various human simulation models are presented in the following texts: *Automotive Ergonomics*, edited by Peacock and Karwowski [1]; *Computer-Aided Ergonomics*, edited by Karwowski, Genaidy, and Asfour [2]; *Simulating Humans: Computer Graphics Animation and Control*, by Badler, Phillips, and Webber [3]; and "Simulating Humans: Ergonomic Analysis in Digital Environments," by Raschke, Schutte and Chaffin [4]. In addition, the keynote addresses by McDaniel at the 1998 SAE Digital Human Conference [5], and by Bubb at the 1999 SAE Digital Human Conference [6] provide excellent overviews of this emerging technology from both an air force perspective [5] and automotive perspective [6].

Perhaps the first attempt to develop a computer simulation of a person performing a reach task was performed by Ryan and Springer for the Boeing Aircraft company in the late '60s [10]. This effort was referred to as the "First Man" program, which later became "Boeman." It was sponsored by the U.S. Naval Air Development Center (NADC) to provide a means of predicting the reach capability of an average size male when seated in a simulated fighter aircraft. It was intended only for in-house development, but was described in a series of public NADC reports in the early '70s. Unfortunately the program used a very inefficient non-linear optimization procedure to predict reach postures, which often took over 20 minutes on a very powerful computer at that time to generate one posture. Needless to say, this did not inspire designers to use it often for "what if" type analyses.

The basic structure of Boeman was adopted in the early '70s by the USAF Aerospace Medical Research Laboratory (AMRL). The AMRL Crew Systems' Interface Division simplified the posture prediction model and added the ability to simulate a variety of male and female anthropometric dimensions while seated in different types of aircraft. The data and algorithms used to refine the reach assessment predictions were developed within the larger

4

Crew Assessment of Reach (CAR) program. The resulting model became known as COMBIMAN. AMRL provided the code to different federal agencies and contractors during the '80s, and continues to maintain and distribute the code. Fig. 1-2 depicts COMBIMAN being used to simulate sight lines in a proposed aircraft. As can be seen, the code uses a simple 3D graphic depiction of the pilot with simple solid forms and shading enhancements.

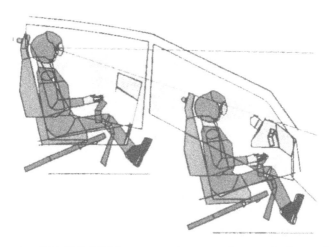

Fig. 1-2 COMBIMAN being used to simulate sight lines for a helicopter crewstation (from McDaniel [7]).

During the early '80s AMRL personnel also recognized a need to simulate aircraft maintenance tasks. This resulted in COMBIMAN being reconfigured to stand, stoop, kneel, and bend while not only reaching about the immediate environment, but also while lifting, pulling, and pushing on various tools and objects placed in the hands. The resulting new model was referred to as CrewChief. Though the reach and sight line analyses provided within CrewChief were a derivative of the algorithms existing in COMBIMAN, the strength functions required totally new population data. These were acquired in a series of empirical studies conducted by the AMRL staff during the '80s which produced an extensive catalog of force predictions which could be accessed through CrewChief. An example of CrewChief rotating a wrench is shown in Fig. 1-3. Both interference in the rotation of the ratchet handle and the population capable of turning the wrench with a given torque can be provided.

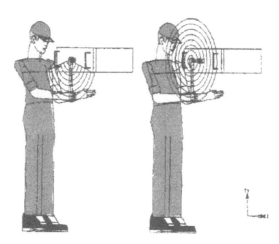

Fig. 1-3 CrewChief is shown rotating a ratchet wrench wherein an obstruction exists on the left but not on the right (from McDaniel [7]).

A more general biomechanical model of population force prediction was developed by Chaffin and his colleagues within the Center for Ergonomics at the University of Michigan in the late '60s. Its logic is fully described in Chaffin et al. [8]. A major initial application of the model was to predict the static strengths of astronauts when lifting, pushing, or pulling on objects during various extravehicular activities in space and on the lunar surface. The model was based on the concept that when a person performs an exertion, the hand forces and body segment weights produce load moments at all the major body joints. By knowing the amount of hand force required to perform a task, the anthropometry of the person, and the posture used by the person, the model can compute these joint load moments. An inverse kinematic algorithm was developed in the late '80s to assist a user in posture predictions. Strength data based on measurements of over 2000 people were acquired by the Michigan group during the '70s, and are used to set limits to the joint load moments. The model also predicts the compressive forces acting on the lumbar spine, which can be compared to limits established by the U.S. National Institute of Occupational Safety and Health (NIOSH). Static body balance and foot slip potential also are predicted by the model. Together these indices allow users to quickly predict the capability of a given population during different manual high exertion tasks of interest. The software is referred as the 3D

Static Strength Prediction Program™, or simply 3DSSPP™. It runs on a personal computer with a Windows operating system, and has been licensed by the University of Michigan to over 2000 organizations and individuals since 1984. A demonstration of this software is shown in Fig. 1-4, wherein a large stock reel is being lifted from the floor.

During the same general period that Boeman and 3DSSPP™ were being developed in the U.S., SAMMIE (System for Aiding Man-Machine Interaction Evaluation) was being developed by Case, Porter, and Bonney at Nottingham and Loughborough Universities in the UK [9]. SAMMIE was conceived as a very general model for assessing various reach, interference, and sight line issues posed by a designer. It uses a sophisticated statistical method to assemble the population

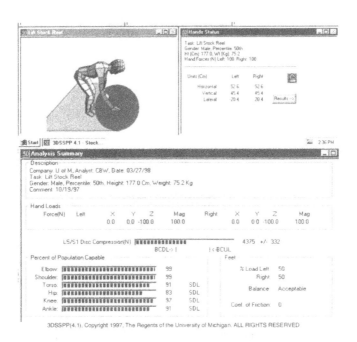

Fig. 1-4 The University of Michigan 3DSSPP™ model shown simulating the lifting of a 200N stock reel from the floor. The analysis indicates that such lifting would be hazardous to the low back (the 4375N low back spinal compression force prediction exceeds the NIOSH 3400N limit) and the exertion also would exceed the hip strength capabilities of about 17 percent of the male population.

anthropometric data needed to predict the percentile size and shape of given somatotype subgroups of interest. Though its human graphic form is shown as a boundary outline, rather than a solid form, it always has been fully integrated into a CAD program which allows users to easily insert the simulated person into a virtual environment of interest, and then visualize the potential effects of various workspace and vehicle designs. Alternative postures can be selected from menus or by direct joint manipulations. Hidden line algorithms allow the figures to appear within a 3D space, as shown in Fig. 1-5. Mirror vision feasibility is provided as an additional feature for vehicle designers. The software is currently available through SAMMIE CAD Ltd. in Loughborough, UK.

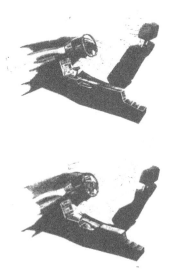

Fig. 1-5 SAMMIE in two different postures to assess combined reach and vision constraints in a proposed vehicle (from Case et al. [9]).

During the '80s many other human simulation models appeared. Only a few will be discussed here, but the reader is referred to the books and papers referenced earlier for brief descriptions of other models, such as ERGOMAN from the Laboratorie d'Anthropologie Appliguée et d'Ecole Hamaine in Paris, WERNER from the Institute fur Arbeitsphysiologie an der Universitat Dortmund, and ANYBODY or its successor ANTHROPOS, which up until recently was marketed by a German firm IST.

For vehicle interior package designs the German model RAMSIS (Realistic Anthropological Mathematical System for Interior Comfort Simulation) deserves special note. It was developed by a consortium of German automobile manufacturers who supported a cooperative arrangement between a software company TECMATH and the Lehrstuhl fur Ergonomie at the Technical University of Munich beginning in 1987. Empirical studies of drivers were undertaken to determine both the postures they chose to use and their psychophysical discomfort when seated for different periods of time in a variety of laboratory seat bucks and vehicles. These data were then combined with a scalable anthropometric model to allow a designer to visualize different size and shape people while driving vehicles having various interior and seat configurations. RAMSIS includes a sophisticated method for representing different population subgroups, and employs an optimization method based on empirical data to predict reasonable postures, thus allowing a designer to easily move the RAMSIS hominoid within a vehicle of interest. The recent hominoid form in RAMSIS uses a fully enfleshed deformable graphic with hidden lines and shadowing to provide a very realistic looking person. Fig. 1-6 depicts the RAMSIS graphic in a contemporary digital mockup of a vehicle as distributed by TECMATH in Germany.

Another more general-purpose model was being developed at the Ecole Politechnique in Montreal, Canada during the '80s. It is now known as SAFEWORK. It incorporates a special statistical model which considers the

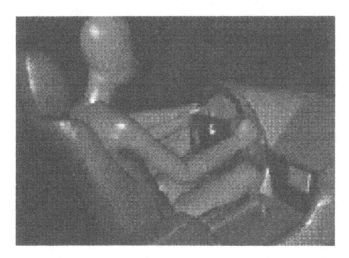

Fig. 1-6 RAMSIS model for vehicle interior design (from Bubb [6]).

multivariate correlation of anthropometric dimensions that define human size and shape. It also has an inverse kinematic method for assisting designers in selecting postures of interest and simulating simple motions. The SAFEWORK hominoid is a fully enfleshed human graphic embedded in a 3D CAD system to render very complicated scenes. Sight lines and physical interference with objects in the virtual environment are provided. Fig. 1-7 depicts the SAFEWORK hominoid. The software is distributed by Genicom Consultants Ltd. in Montreal, Canada.

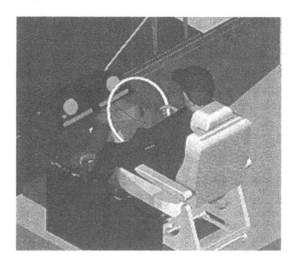

Fig. 1-7 SAFEWORK model shown in workspace evaluation.

One well-known, general-purpose, human simulation model is Jack. It started as a NASA supported effort within the Department of Computer and Information Science at the University of Pennsylvania during the mid-1980s (Badler et al. [3]). It was referred to originally as the TEMPUS model for simulating astronauts assembling the space station. Questions of how an astronaut could best reach and fasten items, as well as see objects while in a spacesuit provided major issues to be considered in the early simulations. Within a period of a few years it was realized that these same issues existed for the operation and maintenance of military aircraft and other ground based vehicles. Jack was then conceived with some of the following features: (1) a means to use different published anthropometric data sets to produce a scalable linkage and hominoid, (2) a flexible spine and multisegmental limbs that could be easily articulated and positioned through an inverse kinematic model, (3) a method for creating a solid form environment in which the Jack hominoid could be easily positioned for reach and visual interference analysis, and (4) a strength guided posture and motion prediction algorithm. Fig. 1-8 depicts a contemporary use of Jack for simulating a new workplace layout. Jack is presently able to be executed on either a PC or graphic workstation, and is continuing to be supported and developed by the Unigraphics Solutions Inc. office in Ann Arbor, Michigan.

Fig. 1-8 Jack being used to simulate a seated workplace.

Because it is so critical to assure that an aircraft pilot can reach to various controls and see different instruments and the outside environment, one aircraft company, McDonnell Douglas (recently acquired by Boeing Aircraft Corporation) embarked on developing its own human simulation system.

11

The Boeing Human Modeling System (BHMS) was first released in 1990 as the Douglas Human Modeling System. BHMS was also known as the McDonnell Douglas Human Modeling System, and more recently the Boeing McDonnell Douglas–Human Modeling System (BMD–HMS).

The BMD–HMS is a tool specifically designed for engineering applications in the aircraft industry. BMD–HMS is a menu-driven, interactive computer program used to define ergonomic design requirements and aid in design evaluation. Engineers at Boeing use BMD–HMS to help visualize and analyze the manual actions required to assemble, maintain, and operate equipment, and to perform quantitative engineering analyses. BMD–HMS provides a set of human modeling and human task simulation tools that allow the user to study human motion and strength; to define design requirements in terms of human reach, vision, and strength capability; and to perform design evaluations in terms of human size accommodation.

BMD–HMS allows the user to create or select the graphic manikin(s) from seven different anthropometric databases, and to select the manikin representation (e.g., nude, clothed, or in a spacesuit) appropriate for the application. Users can also select any of BMD–HMS's applications including, Automated Population Analysis, Reach Accommodation, Static Volume Envelope Generation, Swept Volume Generation, Torque Calculator, Vision Obscuration Plots, and Reach Envelope Generation (the latter is depicted in Fig. 1-9).

1.5 Human Simulation to Enhance Designing Systems for People

The human simulation methods briefly described in the preceding section certainly indicate the existence of a major new design technology that is being rapidly developed by several different sophisticated and well-organized groups. This new technology has the potential to drastically change the process by which most designers decide on the appropriate features needed to improve the interaction of people with the products, tools, and workstations they design.

In an attempt to assess what particular features would be most desirable in future human simulation software, the SAE G-13 Human Modeling Technology and Standards Committee commissioned a survey in 1996 of designers.

Fig. 1-9 The Boeing McDonnell Douglas–Human Modeling System (BMD–HMS) being used to generate a left arm reach envelope for a seated person.

The survey contained a list of almost 500 potential attributes that could be included in future human simulation software. This list was sent to 250 designers around the world with a request that they indicate their preference for each attribute using a scale of 1 (no use) to 5 (highly useful). About 20% responded, most of whom were involved in aerospace applications. These designers showed a strong preference (a score of 4.0 or better) for features such as: (1) being able to include different anthropometric data sets and population demographic subgroups; (2) including a variety of clothing, gloves, and helmets; (3) predicting strength and endurance of different populations in a task; (4) being able to simulate realistic motions and postures with minimum task input descriptions, and in both physically constrained and unconstrained conditions; (5) providing hand grip, strength, and visual sight lines with and without mirrors and obstructions; (6) providing task time line analyses; (7) performing reach and fit analyses for a variety of conditions; and (8) seamlessly accepting I/O commands and data, and/or executing within various CAD systems commonly used in rendering and specifying products, tools, and workstations. The results of this survey were compiled by C. Nelson at Boeing Aircraft and are available by contacting the SAE G-13 Committee.

Though one might criticize this survey as being both superficial in form, and biased by a large proportion of aerospace oriented designers, it is obvious that these designers and engineers wanted to have human simulation software that was easy to use, and that would provide them with valid predictions of a large array of different physical human attributes for diverse populations performing tasks in various and some times extreme conditions. Whether these needs are being met by the existing human simulation programs is a major question in this text. As will be presented in the cases to follow, the answer is not simply stated. In general, the cases indicate that some positive value was realized by using a human simulation program in specific design scenarios, but the type and magnitude of the contribution varies. We'll come back in the final chapter to further discuss this matter.

1.6 Outline of Case Studies and General Issues to Be Presented

The following chapters describe seven design problems in which a human simulation model was employed to assist the designer in determining how best to accommodate a diverse population of people and improve their ability to perform required tasks. The problems addressed include: the placement and size of pedals and the required head clearance in proposed automobiles meant for a global market; the size and placement of handles and steps for safe entry and egress of operators of medium to large size trucks; the specification of window size and control panel locations on the helm of a ship to assure reach and visibility; the accessibility of equipment to be used onboard the International Space Station by both small and large crew members; the location of load stations and baskets for handling of sheet metal parts by two operators in a manufacturing plant to assure safe and effective materials handling; the ability of a small person to reach into a structure to fasten a part; and the design of a trailer to assure safe loading and unloading of munitions onto an aircraft for deployment and maintenance.

Throughout the presentation of these cases it will become clear that available human simulation methods have provided real and important insights to the designers early in the design process, when it still was easy to make changes. Also discussed are issues such as: the time required to use a digital mockup approach for resolving various design questions, as opposed to creating a physical prototype and testing it with volunteers; the type of data structures needed to create a digital mockup and perform the needed human simulations; the validity and/or ability to verify the human simulation results; and the ability of the resulting simulations to be communicated effectively to those most affected by the proposed designs.

Clearly seven case studies are insufficient to substantiate strong conclusions about the relative merits and limitations of the human simulation programs that are now being used. It is believed, however, that collectively these cases provide meaningful insights as to how this rapidly emerging technology will eventually affect the design of all products, tools, and workplaces in the future.

References

1. Peacock, B. and Karwowski, W., *Automotive Ergonomics*, Taylor & Francis, Washington, D.C., 1993.
2. Karwowski, W.; Genaidy, A.M.; and Asfour, S.S. (Eds.), *Computer-Aided Ergonomics*, pp. 138–156, Taylor & Francis, London, 1990.
3. Badler, N.I.; Phillips, C.B.; and Webber, B.L., *Simulating Humans: Computer Graphics Animation and Control*, Oxford University Press, New York, 1993.
4. Raschke, U.; Schutte, L.; and Chaffin, D.B., "Simulating Humans: Ergonomic Analysis in Digital Environments," in Salvendy, G. (Ed.), *Handbook of Industrial Engineering*, J. Wiley & Sons, New York, 2001.
5. McDaniel, J.W., "Human Modeling: Yesterday, Today, and Tomorrow," SAE International Conference, Digital Human Modeling for Design and Engineering, Dayton, Ohio, 1998.
6. Bubb, H., "Human Modelling in the Past and Future—The Lines of European Development," SAE International Human Modeling for Design and Engineering Conference—The Hague, The Netherlands, 1999.
7. McDaniel, J.W., "Models for Ergonomic Analysis and Design: COMBIMAN and CREWCHIEF," in Karwowski, W.; Genaidy, A.M.; and Asfour, S.S. (Eds.), *Computer-Aided Ergonomics*, pp. 138–156, Taylor & Francis, London, 1990.
8. Chaffin, D.B.; Andersson, G.B.J.; and Martin, B.J., *Occupational Biomechanics, 3rd Edition*, John Wiley & Sons, New York, 1999.

9. Case, K.; Porter, J.M.; and Bonney, M.C., "SAMMIE: A Man and Workplace Modeling System," in Karwowski, W.; Genaidy, A.M.; and Asfour, S.S. (Eds.), *Computer-Aided Ergonomics*, p. 47, Taylor & Francis, London, 1990.

10. Ryan, P.W. and Springer, W.E., "Cockpit Geometry Evaluation Final Report," Vol. V, JANAIR Report 69105, Office of Naval Research, Washington, D.C., 1969.

Chapter 2

Anthropometric Analyses of Crew Interfaces and Component Accessibility for the International Space Station

Cynthia Nelson, CPE
The Boeing Company

2.1 Introduction

This chapter describes the human modeling simulation that demonstrated compliance with contractual specifications for the International Space Station (ISS). Anthropometric analyses of ISS Node 1 on-orbit crew interfaces were performed using the Boeing Human Modeling System. Human simulation was used to verify and document that Node 1 on-orbit replaceable units are accessible by the smallest and largest crewmembers. Human simulation was also used to analyze and verify that normal operation hardware and emergency use hardware are compatible with anthropometric requirements. Boeing satisfied contractual requirements and realized substantial cost and schedule savings by replacing expensive mockup, test hardware, and flight hardware testing with electronic human modeling simulation.

A specific example of each of the three types of simulated intravehicular activity tasks (i.e., normal or routine, maintenance, and emergency) is described. The maintenance task example describes the job simulated (including reference to the source Logistics Support Analysis procedure and specified hand-tools) and the results of the task simulation. Similar details are provided for the normal and emergency task descriptions. Each example describes how human simulation was used to verify and document compliance with physical and visual accommodation of each crewmember performing the task, and tool clearance specifications. Based on the experience gained through ISS Node 1 task simulations, human modeling system limitations are discussed and recommendations for improving this technology are made for these types of applications.

2.2 International Space Station Node 1

Node 1, or Unity Node, is the first Space Station node and the first major U.S.-built component of the station [1]. It was delivered by the space shuttle in December of 1998 [2]. The shuttle crew conducted three spacewalks to attach Unity to the Russian-built control module, Zarya.

Unity Node is a connecting passageway to living and work areas of the ISS. It has six hatches that serve as docking ports for the other modules including the lab and habitat modules. Node 1 is 18 feet long and 15 feet in diameter with four equipment racks. Unity is fabricated of aluminum; it contains 216 lines to carry fluids and gasses, 121 internal and external electrical cables using six miles of wire, and more than 500,000 mechanical items.

2.3 Requirements for Anthropometric Accommodation and Component Accessibility

The design of Node 1 must provide crewmembers with physical and visual access to all components required for normal, maintenance, and emergency operations. The configuration and layout of Node 1 internal hardware was designed to enable a wide range of body sizes to operate, maintain, and perform emergency operations. Our customer specified that Node 1 internal hardware must accommodate the 5th percentile Japanese female to the 95th percentile American male anthropometric size measurements [3]. Anthropometric analyses of all on-orbit crew interfaces had to be performed to verify that Node 1 hardware accommodates the specified range of anthropometric sizes. The location and installation of all on-orbit replaceable units (ORUs) also had to facilitate component removal and replacement, and meet tool clearance specifications.

2.4 Human Modeling Simulation versus Traditional Anthropometric Testing

Boeing chose to use human modeling simulation and digital mockups (DMUs) to perform the required anthropometric analyses and component accessibility demonstration of Node 1 hardware. We could have used more traditional physical devices such as mockups, test hardware, or flight hardware; however, utilization of DMUs and human modeling technology offered several benefits over traditional anthropometric testing.

Electronic simulation enables us to analyze our prototype designs and permits our analyses to occur earlier in the design process than traditional testing. Any problems we uncover during DMU evaluation, before hardware is built, are less costly to rectify than are problems uncovered during traditional testing with mockups or hardware built from detailed drawings. By using electronic simulation rather than traditional testing we can reduce costs associated with rework and scrappage, and risks associated with customer disapproval and schedule slippage.

Traditional anthropometric testing requires expensive physical structures (e.g., mockups, prototypes, or completed hardware) and human subjects. Human modeling requires access to computer aided design (CAD) drawings and appropriate-sized human manikins. By replacing expensive physical devices with digital mockups, and human subjects with manikins, cost and schedule savings can be realized.

Because anthropometric analysis of Node 1 was the first time Boeing applied human modeling on a large scale, we could not realistically estimate either cost or schedule savings. However, we did plan to document the savings of using DMU evaluation over traditional anthropometric testing. We used the Boeing Human Modeling System (BHMS) [4] to simulate and analyze crew interfaces and component accessibility.

19

2.5 Anthropometric Accommodation

BHMS was used to verify and document that the specified range of anthropometric sizes could perform all normal, maintenance, and emergency tasks. All crew interfaces were analyzed to verify that the 5th percentile Japanese female functional reach limits were not exceeded and the 95th percentile American male body envelope was not confined.

Human simulation was used to demonstrate that all Node 1 hardware is accessible by the smallest and largest crewmembers. BHMS's Create Manikin function and dimensional data from NASA's anthropometric database were used to create the two manikins. These two manikins represent the worst cases or extreme individuals that must be accommodated.

The input dimensions used to create the larger manikin were based on anthropometric data for the 95th percentile American male; the dimensions were adjusted to account for spinal lengthening due to weightlessness. Full spinal lengthening (i.e., approximately 3 percent "growth" in height) occurs within the first two days of weightlessness. This manikin represents the largest male crewmember after he has experienced full spinal lengthening; his 6'2" stature in 1-G lengthens to approximately 6'5" in a 0-G environment.

The input dimensions used to create the smaller manikin were based on anthropometric data for the 5th percentile Japanese female. Unlike the larger male manikin, the female manikin's dimensions were not adjusted to account for spinal lengthening due to weightlessness. The shortest functional reach would be that of a small female crewmember who had just recently arrived onboard the ISS and had not yet experienced any spinal lengthening. Therefore, the input dimensions used to create this manikin were 1-G measurements. This manikin represents the smallest crewmember before she has experienced the effects of microgravity spinal lengthening. Her stature is the same as if she were in 1-G (approximately 4'10").

It is important to note that before any modeling system can be used to demonstrate anthropometric accommodation, it must be validated. BHMS had been validated before Node 1 analyses were performed. The BHMS Validation Manual [5] documents the extent to which BHMS accurately represents the physical human body. BHMS validation results include: (a) a mean error of

0.2 percent in overhead fingertip reach, sitting, and (b) a mean error of less than 4.2 percent between any manikin measurement and the corresponding input dimension.

Compliance of BHMS to NASA requirements for body sizing and body joint limits had also been established before Node 1 analysis. Boeing validated and documented the BHMS intravehicular activity (IVA) task analysis process [6], and documented compliance of the BHMS manikin to NASA's flight crew integration standard requirements [7].

2.6 Anthropometric Analyses and Component Accessibility Demonstration with BHMS

BHMS was used to simulate and analyze all Node 1 IVA tasks. Three hundred seventy IVA tasks were analyzed including 42 normal operations, 252 maintenance operations, and 76 emergency operations. A specific example of each of the three types of tasks is described below. Each example describes the task simulated and the results of the analysis. In addition, each example describes how human simulation was used to verify and document compliance with physical and visual accommodation of each crewmember performing the task, and tool clearance specifications.

2.7 Normal Operation—Damper Valve Control Assembly

Simulation of normal operations was relatively straightforward: import appropriate Node 1 CAD drawings into BHMS, assemble drawings into a DMU, place a tag point on the control, select a manikin, attach and position the manikin at a handrail, and perform a reach to the control's tag point.

The initial posture we used to attach the manikin within the DMU was a user-defined neutral body posture (NBP) modified for manikin attachment by the hand. The NBP is the relaxed, unrestrained posture the human automatically assumes in weightlessness. We tried to keep manikin postures as close as possible to the NBP since discomfort and fatigue can result if the body is required to assume other postures in 0-G. However, we usually had to create iterative body postures that deviated from the NBP in order for the manikin to

21

obtain visual and physical access to the control or component while attached to a handrail. We took advantage of BHMS's Single Reach function to help position and posture the manikins. This feature allowed us to put the manikin in our initial NBP, constrain the hand to the handrail, and reach to the control point. If the control could not be reached, we then modified the position of the arm that was constrained to the handrail, and/or we moved the handrail attach point to enable physical access of the control.

Another posturing approach we used was to create an eye reference point (ERP) from which the manikin could gain visual access to the control or component. We then attached the manikin at the ERP and used the Single Reach function to move one hand to the control and the other hand to the handrail. It was beneficial to have both manikin posturing methods available.

It is important to note that BHMS's joint limits file contains appropriate and realistic ranges of motion for each joint. This joint limits file ensures that BHMS manikin postures and reaches approximate the body postures and reaches that real crewmembers can realistically achieve.

Normal operation of the damper valve control assembly (i.e., the temperature control switch) by the smallest crewmember is depicted in Fig. 2-1. This simulation demonstrated that the location and orientation of the switch accommodates both the smallest and largest crewmembers; visual and physical access to the temperature control switch can be achieved.

While simulation of normal operations was relatively straightforward, it was often a challenge to determine which of the 20 handrails to use and where to place the manikin anchor point on the selected handrail.

The major source of frustration we experienced in simulating normal operations was the time-consuming and tedious work required building the DMU; this was also a problem with maintenance and emergency task simulation. Before the DMU could be assembled, the appropriate electronic drawings had to be located in the host CAD system. These drawings then had to be translated and the data transferred to the workstation on which BHMS was hosted. Once the DMU was assembled, user productivity was usually not optimal due to the negative effects of the large DMU size upon the speed of the system. While building a DMU is time-consuming, it is still faster to build

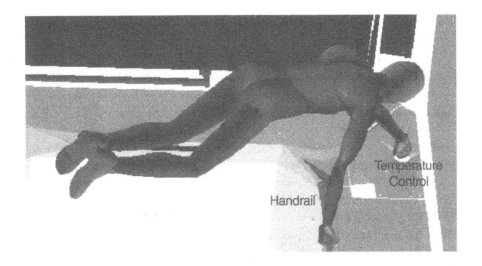

Fig. 2-1 Physical access to temperature control by small female (using handrail as restraint). (Courtesy of The Boeing Company.)

a DMU than to build a physical mockup. DMU construction is also less costly than building a physical mockup; DMUs require significantly less equipment and facilities than do physical mockups. A DMU requires a desktop computer workstation, while a mockup requires a large room or hangar space, and physical construction materials.

2.8 Maintenance Operation—Inter-Module Venting Fan Assembly

Simulation of maintenance operations was more complicated than simulation of either normal or emergency operations. This complexity was due primarily to the fact that maintenance operations involve multiple subtasks, hand-tool utilization, and restraint of the body by a footplate.

The absence of gravitational forces leaves crewmembers without any stabilization when they exert force with a hand-tool; therefore, a body restraint system is required. Foot restraints provide crewmembers with good reach performance, stability, and control needed for maintenance operations.

Simulation of most Node 1 maintenance tasks required the manikin to be attached at the footplate.

The footplate was located along the length of one of the handrails and then rotated such that the manikin would be aligned with the on-orbit replaceable unit (ORU). The manikin was then attached to the footplate. If the location of the footplate did not afford visual or physical access to the ORU, then the footplate had to be relocated and/or repositioned. This iterative process of manikin attach was often time-consuming. In addition, it was often a challenge to determine which handrail and footplate location combination to use. The handrail/footplate combination had to provide adequate physical access to the ORU while maintaining visual access, a manikin body posture similar to the neutral body posture (NBP), and enable a reach that did not max out shoulder joint limits. We often found that the handrail/footplate combination that worked for the larger male manikin would not provide the smaller female manikin with adequate access to the ORU; in this case, each manikin required a unique handrail/footplate combination.

The maintenance task example that follows provides "snapshots" of the major subtasks required to access the inter-module venting fan assembly for removal and replacement. This simulation was based on the Logistics Support Analysis (LSA) procedure that provides maintenance details including sequence of steps, component(s) to be accessed, type and number of fasteners, and hand-tool(s) required.

2.8.1 Maintenance Subtask 1—Removal of Closeout Panels

The first step in this maintenance operation is to remove three closeout panels in order to gain access to the fan assembly. Removal of all three panels requires the crewmember to unfasten 29 captive fasteners. Fig. 2-2 shows the smaller female crewmember restrained by the footplate reaching to unfasten a single captive fastener on one of the closeout panels using a hex driver and handle. Simulation of this subtask demonstrated that both the smallest and largest crewmembers have sufficient visual and physical access to remove all three panels.

Fig. 2-2 Small female (restrained by footplate) unfastening fastener #4 on closeout panel NOD104-03. (Courtesy of The Boeing Company.)

2.8.2 Maintenance Subtask 2—Removal of Band Clamps

The second major subtask is to remove the band clamps securing the coupling to the lower silencer and the fan assembly. Fig. 2-3 depicts the larger male crewmember using a socket and handle to unfasten the band clamp securing the upper coupling to the fan. Fig. 2-4 shows the crewmember's partially obstructed visual access of the band clamp, and Fig. 2-5 shows acceptable access for hand-tool actuation. Simulation of this subtask involved many more steps than those illustrated in Figs. 2-3 through 2-5. Human modeling enabled us to demonstrate that there is adequate access for hand-tool actuation; the band clamps can be unfastened despite the limited visibility.

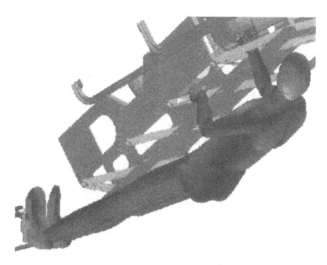

Fig. 2-3 Large male (restrained by footplate) unfastening band clamp securing upper coupling to fan. (Courtesy of The Boeing Company.)

Fig. 2-4 Partially obstructed vision of upper band clamp by large male (view from eye point in posture shown in Fig. 2-3). (Courtesy of The Boeing Company.)

Fig. 2-5 Access for hand-tool actuation by large male at upper band clamp. (Courtesy of The Boeing Company.)

2.8.3 Maintenance Subtask 3—Removal of Lower Silencer

The third subtask in this maintenance operation is to remove the lower silencer; removal of the lower silencer is required to enable physical removal of the fan assembly. Four captive fasteners securing the silencer to the mounting bracket must be removed with a hex driver and handle. Fig. 2-6 depicts one of several body postures the small female crewmember may assume for removal of a fastener securing the silencer. Fig. 2-7 depicts reach to the same fastener by the large male crewmember with the hand-tool. Electronic simulation of this subtask demonstrated that both the smallest and largest crewmembers have adequate visual and physical access to remove the lower silencer.

Fig. 2-6 Small female (restrained by footplate) unfastening #1 fastener securing lower silencer with one hand. (Courtesy of The Boeing Company.)

Fig. 2-7 Large male (restrained by footplate) unfastening #1 fastener securing lower silencer. (Courtesy of The Boeing Company.)

2.8.4 Maintenance Subtask 4—Removal of Fan Assembly

The final subtask is removal of the inter-module venting fan assembly. Removal of this ORU requires the crewmember to unfasten four captive fasteners securing the fan to the mounting plate. Fig. 2-8 shows physical access to one of the fasteners by the small female crewmember. The tool clearance envelope is the volume around the fastener head that should be free in order to provide crewmembers with optimal physical clearance for tool actuation. Each hand-tool has a unique tool clearance envelope. Simulation of this subtask demonstrated that the tool can be actuated (see Fig. 2-9) by both the largest and smallest crewmembers, although there is interference with the structure and the fan by the tool clearance envelope (see Fig. 2-10).

Fig. 2-8 Small female (restrained by footplate) unfastening #4 fastener securing fan. (Courtesy of The Boeing Company.)

29

Fig. 2-9 Tool actuation of #4 fastener securing fan.
(Courtesy of The Boeing Company.)

Fig. 2-10 Tool clearance envelope for #4 fastener securing fan.
(Note interference with structure and fan.)
(Courtesy of The Boeing Company.)

2.8.5 Maintenance Task Summary

BHMS enabled us to analyze the Logistics Support Analysis (LSA) procedure for removal and replacement of the fan assembly. We were able to verify task sequencing, access behind closeout panels, hand-tool clearances, and component accessibility before any hardware was built. This simulation also demonstrated that both the smallest and the largest crewmember could accomplish removal and replacement of the inter-module venting fan assembly.

2.9 Emergency Operation—Portable Fire Extinguisher at Fire Port

We also used BHMS and Node 1 DMUs to demonstrate that both the smallest and largest crewmembers could retrieve and activate emergency equipment. Fig. 2-11 depicts physical access to the emergency locker for retrieval of the portable fire extinguisher by the largest crewmember. Fig. 2-12 shows the subsequent application of the extinguishing agent into one of the 26 fire ports in Node 1. This simulation demonstrated that the emergency operation could successfully be accomplished.

Fig. 2-11 Physical access to emergency locker by large male (using handrail as restraint). (Courtesy of The Boeing Company.)

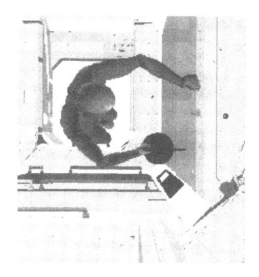

Fig. 2-12 Large male (using handrail as restraint) applying extinguishing agent to fire port NOD1P3-1. (Graphic by Charles Gidcumb. Courtesy of The Boeing Company.)

2.10 Compliance Demonstration and Simulation Benefits

Boeing satisfied contractual requirements and realized substantial cost savings (greater than 1 million dollars) by replacing expensive mockup, test hardware, and flight hardware testing with BHMS electronic simulation. Human modeling was used to verify visual and physical access to all required controls, fasteners, and components, and compliance with tool clearance requirements. Human performance factors such as force requirements and strength capabilities were not considered during BHMS analysis of Node 1.

Human simulation was used successfully to demonstrate that Node 1 hardware is accessible; all required tasks can be performed by the smallest and largest crewmember [8]. In addition to the cost savings, demonstration of component accessibility, and verification of anthropometric accommodation there were other benefits of Node 1 task simulation.

Human simulation enabled Boeing to review and revise LSA procedures before any hardware was built. Boeing engineers were able to check that the access afforded to the ORU with the removal of the specified closeout panel(s) was adequate. We were able to verify that the hand-tool specified for any given maintenance task could be used in the working space. Maintenance simulation allowed us to "perform" the remove and replace task. We could then alter the procedure, check access through various closeout panels, and try different hand-tools to optimize ease of maintenance.

We received additional benefits from Node 1 task simulations. The graphical images output from the human modeling simulations enhanced documentation and facilitated communication. We used these images to document compliance with contractual specifications. This documentation format also allowed us to post our results on the web [9]. Intranet distribution of program documentation facilitated discussion within Boeing and communication with our customer.

The benefits realized with Node 1 simulation have lead to Boeing and NASA acceptance of human modeling technology as a viable design and analysis tool.

2.11 Recommendations

While there are many benefits associated with human modeling simulation, there is still plenty of room for improvement. Human modeling technology will have a wider appeal and a better utilization rate as BHMS enhancements continue.

2.11.1 Improvements for Microgravity Simulation

There are several unique human modeling requirements to consider for microgravity simulation. Human modeling systems should automatically recalculate or adjust those anthropometric dimensions affected by microgravity. The user should be able to create a manikin based on 1-G dimensions and the modeling system should automatically adjust the manikin for spinal lengthening. Crewmembers cannot maintain a 1-G posture in a microgravity environment; therefore, reach and motion algorithms should take into account the exertion required to deviate from the neutral body posture. A simple, easy-to-use user interface should provide for manikin attachment to a foot restraint and rapid repositioning of both the manikin and the footplate.

Since completion of Node 1 analysis, the Boeing BHMS development team has implemented a toggle to automatically adjust manikin dimensions for either a 0-G or 1-G environment (i.e., with or without spinal lengthening due to weightlessness). New reach and motion algorithms and an improved manikin-positioning interface are in work.

IVA and extravehicular activity (EVA) task analyses will be more complete when human limitations and capabilities such as strength and postural comfort are considered (in addition to visual and physical accommodation). System operability and maintainability can be enhanced during the integrated product development process by taking into account human capabilities and limitations of the people who are expected to assemble, operate, and maintain the system. Boeing is continually upgrading BHMS; currently we are focusing on strength capability.

2.11.2 Improvements for User Productivity

Seamless integration between the human modeling system and the host CAD system is critical for improving user productivity and reducing schedule time. The requirement for CAD drawings to be imported into the human modeling system is very time-consuming. Simulation set-up time would drop dramatically if CAD drawings did not need to be translated and imported into the human modeling system. Since completion of Node 1 analysis, BHMS enhancements (e.g., a VRML [virtual reality modeling language] translator for 3D file interchange) for easier and faster construction of DMUs have been made; future BHMS and CAD improvements are planned.

Additional BHMS improvements will increase user productivity and help broaden the appeal of human modeling technology. One improvement is a platform independent video/image capture capability to support documentation and training efforts. Another improvement is an enhanced user interface with wizards. A wizard is a feature that asks questions and then uses the answers to automatically perform a function, such as manikin attach. Wizards will help new users perform simulations and improve user productivity in general.

2.12 Conclusions

This chapter describes how human modeling technology has advanced at Boeing to become a mainstream design validation tool, and how it was used to demonstrate compliance with contractual specifications for ISS Node 1 internal hardware. Human simulation enabled Boeing to analyze crew interfaces and verify Node 1 design in a cost-efficient and timely manner. Boeing was able to demonstrate that Node 1 is compliant with anthropometric requirements based on these analyses.

Human simulation is still a relatively young technology with room for improvement. However, the benefits and savings Boeing and NASA realized with Node 1 analyses ensure the future of human modeling technology for IVA analysis. Boeing is using human modeling technology to analyze and demonstrate anthropometric accommodation for both the ISS lab and propulsion modules.

References

1. http://spaceflight.nasa.gov/station/assembly/elements/node1/
2. http://spaceflight.nasa.gov/station/assembly/flights/2a.html
3. National Aeronautics and Space Administration, *International Space Station Flight Crew Integration Standard*, NASA-STD-3000/T Rev B, Space Station Program Office - Johnson Space Center, Houston, Tex., 1995.
4. http://www.boeing.com/assocproducts/hms/
5. Nelson, C.A., *The Boeing McDonnell Douglas Human Modeling System: Validation Manual*, Tech. Report MDC 97K0074, Rev. A, The Boeing Company, Long Beach, Calif., April 1998.
6. Peterson, P.J., "Compliance of McDonnell Douglas Human Modeling System (MDHMS) Manikins to SSP 50005B," Tech. Report MDC 97H0579, McDonnell Douglas Corporation, Huntington Beach, Calif., May 1997.
7. Peterson, P.J., "Analytical Model Control and Documentation of the McDonnell Douglas Human Modeling System (MDHMS)," Tech. Report MDC 96H0614, McDonnell Douglas Corporation, Huntington Beach, Calif., September 1996.

8. Peterson, P.J., "Node 1 Anthropometric Human Engineering Verification Analysis," Tech. Report MDC 97H0490, The Boeing Company, Huntington Beach, Calif., April, 1998.
9. http://iss-www.jsc.nasa.gov/issapt/iva/node1/490/490.html

Chapter 3

Human Model Evaluations of Air Force System Designs

John D. Ianni
AFRL/HECP

3.1 Introduction

This chapter describes how human modeling has been used in the evaluation of three Air Force equipment designs. This includes maintainability of F-15 radar components, replacement of an F-22 power supply, and general ergonomics of a new munitions trailer. These three evaluations have similarities as well as interesting differences, which will be discussed. The steps used in each analysis are typical: acquire the system models, integrate human models, create the task simulations, and perform the analyses. All of these steps will be described for the munitions trailer analysis, but, to limit redundancy, only highlights of the F-15 and F-22 analyses will be presented. After discussing these specific analyses, I will share my views of this technology, with reference to these case studies, as well as other experiences with human modeling.

3.2 Background

The Air Force and other segments of the Department of Defense are using new technology to migrate away from the need for on-board operators or pilots. The new uninhabited vehicles will, to a large extent, operate autonomously with minimal human control. This is a logical move if one considers the high value of human life, the design requirements imposed for life support, and the cost of highly trained operators. However, despite this trend, there will continue to be a human-in-the-loop for all systems because people must physically interact with the vehicle at least during manufacturing, deployment, and sustainment. Sustainment, which involves maintaining or servicing a system to ensure that it remains operational, will continue to be a human-intensive activity. As a matter of fact, an estimated 30% of life cycle costs have traditionally been related to maintenance, and a major portion of this is personnel costs. These costs can be reduced in some cases through better equipment designs [1].

Most maintenance technicians whom I have interviewed state that design oversights can lengthen maintenance procedures manyfold. What could be a simple removal of a high-failure component is often complicated by other components that hinder accessibility, so removing one bad component can require removing several good components. These good components in turn fail more often, since they must be repeatedly disconnected and reconnected just for accessibility. This amounts to a huge expense over the long life of Air Force systems. For this reason and others, computer-based human figure models are emerging as an invaluable tool in the analysis of Air Force system designs. As discussed in later, taking extra time to simulate human interaction with equipment can pay off greatly over the system's life.

In this chapter, I will refer to computer-based human models, human figure models, manikins, virtual humans, digital humans, and avatars pretty much interchangeably, but there are some distinctions that should be made. A manikin, the most basic form of human model, can be put into poses, but typically does not simulate movement. Virtual or digital humans, on the other hand, are able to simulate movement and task performance. They may even include some reasoning about task performance and be able to handle variances in the environment. An avatar is a moving representation of a real person tracked with virtual reality equipment. These distinctions are important because the type of human model used greatly impacts the way the model is used for ergonomic assessments.

Before discussing the case studies, let me discuss my previous experiences with human modeling. I have spent the better part of my career developing, promoting, and using human models to ensure that maintainability is adequately considered before a system is built. Starting in 1985, I was involved in the development of a 3D computer-aided design (CAD) manikin called Crew-Chief. CrewChief was a static (i.e., motionless) human model that interfaced directly to various CAD systems. Because it ran native to the CAD environment, no geometric data conversion was necessary and the user interface was familiar to users of that system. The software provided strength information based on gender and posture, visibility plots, and collision predictions for reach assessment. In many circumstances CrewChief's capabilities were sufficient, but the lack of simulated motion made it insufficient for complete task analysis.

In 1991, the Air Force started the DEPTH (Design Evaluation for Personnel, Training, and Human Factors) program to add maintenance analysis capabilities to an articulated figure modeling system called Jack [2]. Jack, then under development at the University of Pennsylvania, was one of the most advanced human modeling systems, with features such as a detailed spine and real-time joint manipulation. The DEPTH program enhanced existing Jack features and added maintenance analysis functionality external to the Jack software, much of which was later integrated into the Engineering Animation Incorporated (EAI) commercial version of Jack now supported by Unigraphics Solutions Incorporated [3].

In addition to developing functionality for maintenance analysis on the DEPTH program, some real-world demonstrations were conducted. Two of these demonstrations, F-15 radar maintenance and F-22 power supply replacement, are discussed in the following sections. Although these analyses were touted as "demonstrations," they were actually useful analyses of new (F-22) and modified (F-15 radar) systems [2].

3.3 F-15 Radar Analysis

The F-15 analysis, performed for the Air Force Research Laboratory (AFRL) by Hughes Missile Systems Company in 1997, involved several upgraded radar components that were arranged differently from the original design. This arrangement affected the inspection and maintenance procedures, so it was thought to be a good candidate to demonstrate DEPTH's analysis capabilities. The procedures simulated in this effort, graphically depicted in Fig. 3-1, are described below.

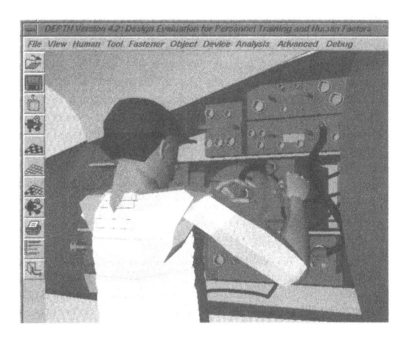

Fig. 3-1 DEPTH simulation of connector removals on F-15 radar components.

1. A maintenance technician opened the radome (the nose of the aircraft). A small-stature human was initially used to demonstrate an unsuccessful reach. As expected, the simulation reported that the technician failed to reach the radome. After a human of average stature was substituted, the simulation showed that the task could be accomplished successfully.

2. Next, the radome was secured with the lockout bar. This illustrated object handling from the technician's viewpoint. The original plan was to perform the visual search for the lockout bar and show that the view was obstructed. This was not possible due to a problem with the human modeling software. Since the collision list was used for visibility and collision checking, the "adjust joint" function reported erroneous collisions. However, the proper placement of the lockout bar was demonstrated.

3. A visual inspection of the radar antenna was conducted next. Again, this was simulated through the technician's eyes.

4. Next, the analog signal converter was disconnected. This activity was intended to demonstrate collision detection when oversized connectors were substituted in the environment. However, the collision detection could not be performed because user-defined motions could not access the collision list. Modifying the software to provide this capability was not possible given time constraints; however disconnecting the analog signal converter enabled the removal of cables and fasteners.

5. The analog signal converter was removed and replaced from the aircraft bay. This demonstrated the ability of one person to pull and lower an object. Since the mass of the component exceeded the weight limits for one person, the ability to detect and present a strength warning was demonstrated with a dialog box and a warning icon at the bottom of the DEPTH window. This was calculated using the task-based strength algorithms from the CrewChief software. Exceeding the strength limitation was not a surprise, since the component was similar in size and weight to the original design. For demonstration purposes, we adjusted the weight (mass) of the component and reran the simulation. This time the human model was able to perform removal and reinstallation, including the reconnection of cables and fasteners. Similarly, the transmitter disconnection, removal, and replacement demonstrated a two-man push, pull, and lift [2].

Since this analysis was performed in conjunction with DEPTH development, it took nearly a year for one person part-time to complete, including time to track down CAD data and model additional equipment. Although a significant number of CAD files were located, a considerable amount of time was spent modeling moving parts such as connectors, cables, fasteners, wave-guides, and the B4 multipurpose aircraft maintenance stand. This crank-raised stand provided a working platform during the removal of the analog signal converter. In addition the analysis process was stalled because some functionality identified during the analysis had to be developed.

Fortunately for the radar designers, no unexpected maintenance problems were identified. However, the simulations were significant because they demonstrated DEPTH's new task simulation capability. This capability, collectively called motion models, allowed fairly complex tasks to be automated using a hierarchy of subtasks. This goal-driven approach provided advantages from previous scripted or animation-based simulations [10]. First, it allowed the user to state *what* simulations were desired rather than *how* they were to be done. Second, it gave the human modeling system an "understanding" of the abstract task concept, which could be useful in creating a natural language description of the task for technical manuals. Third, it allowed task times to be calculated automatically from times associated with each task primitive [5].

On the negative side, the motion models did not always simulate tasks in realistic ways. For example, if instructed to walk to a certain site, the human model would start a seemingly logical approach but then walk past the target as if it were not seen. Sometimes the human model would walk in a continuous loop around the target. In other cases, if the human was instructed to reach for certain connectors, the reach would be performed with unnatural arm movements, sometimes declaring obviously simple reaches unsuccessful. Despite these problems, which had to be worked around, the motion models successfully demonstrated the benefits of autonomous task performance. As discussed in Section 3.8, this autonomy will be key if digital humans are to be used by a wider audience.

3.4 F-22 Power Supply Analysis

The F-22 analysis, shown in its final version in Fig. 3-2, came about as a result of a request from the Air Force's F-22 System Program Office. The evaluation, performed for AFRL by Lockheed-Martin Tactical Aircraft Systems, involved the removal and replacement of the left-forward electronic warfare power supply line replaceable unit. This unit had questionable accessibility, given its location, orientation, and obscuration by other equipment. To create the detailed virtual mockup of the power supply mounting area, both existing CAD geometry and Jack's modeling functionality were used. The geometry imported from the CAD system (Dassault Systems' CATIA®) was modified to remove extraneous information that would bog down the computer. Missing geometric and mechanical information was modeled based on paper-based schematics.

The results of this evaluation were more insightful than those from the F-15 evaluation. The first step involved removing some electrical connectors on the

Fig. 3-2 DEPTH F-22 Analysis: Human is facing power supply, which is not visible in this picture.

rear of the power supply. This task proved to be difficult for short females because a relatively high and deep reach was required. Even with a ladder, the woman would risk losing her balance as she reached into the bay. Large males also encountered problems on this task due to limited hand space (Fig. 3-3). The second step involved removing the fasteners, which was not problematic for the populations tested. The final steps involved lifting the front of the unit up with one hand, slipping the other hand under the unit, and raising the unit out of the bay. Lockheed-Martin analysts found that a small female did not have sufficient accessibility or the strength necessary to safely perform these steps, even with a handle on the component. The simulation showed that the power supply would be near her maximum reach and strength, and therefore she may not be able to control the unit safely. Without this control, the power supply could fall into the maintainer's face. Therefore it was recommended that smaller or weaker individuals not perform the task alone and that a handle on the front of the unit would be necessary for others.

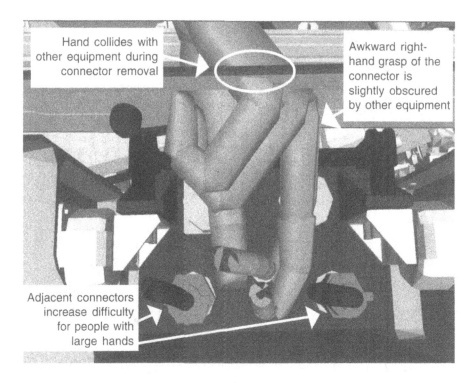

Fig. 3-3 Difficult accessibility for large male hand behind the F-22 power supply.

None of the steps in putting the F-15 and F-22 demonstrations together were trivial. The most difficult task, however, was tracking down the CAD information. This was partly because the prime contractors did not have all of their subcontractor's CAD data. Even when the CAD data were delivered, they were often found to have been created with a different CAD system. Thus it is not surprising that few, if any, weapon systems have a complete set of 3D electronic design information readily accessible, let alone integrated. This is not to imply negligence on the part of the prime contractor, because it can be prohibitively costly (and possibly a security breach) to consolidate all of the latest CAD information of an entire weapon system. This situation will hopefully change with CAD data standardization and improved data security techniques.

3.5 Munitions Trailer Analysis

When the Air Expeditionary Force (AEF) originally asked my organization to analyze human factors on the Next Generation Munitions Trailer (NGMT), there was a team for DEPTH development who worked with human modeling on a daily basis. We had a complete multidisciplinary team of human factors engineers, modelers, and simulation specialists on hand who could spare an hour here or there to help in this analysis. But when AEF was actually ready to start the analysis, the DEPTH contract had finished and the team was no longer in place. It was suggested that AEF perform the analysis at their site since they were most familiar with the trailer. But AEF had no personnel available who could do the task and, even if they had, it would have taken considerable time to set up the hardware, software, and training at their facility in Mountain Home, Idaho.

Therefore, if an ergonomic analysis of the munitions trailer was going to be performed, it would be up to me. At first this did not seem to be a problem since the trailer was relatively simple both geometrically and mechanically. However the analysis took longer than expected—over four months part-time. Although I was experienced in human modeling and simulation, there was a learning curve with the domain area (munitions handling), model creation, and the new version of Jack. Human factors analyses should ideally be performed or supervised by someone who is trained in ergonomics, but as in this case, they often are not. The people performing these analyses can be ex-maintenance technicians, design engineers, CAD operators, or even scientists in a laboratory.

This creates a great challenge for human model developers because few prerequisites can be assumed. The software needs to be simple enough for practically anyone to use. However, due to the complexity of this type of simulation, none of the software tools that I know of can be correctly used without understanding 3D computer graphics, simulation concepts, and human factors engineering. Automating task performance is one step that can be taken to make this happen [6].

3.6 Modeling the Trailer

The NGMT (physical prototype shown in Fig. 3-4) is a relatively simple, extendable trailer which can handle many types of munitions. The bed of the trailer can be expanded much like a dining room table to accommodate most sizes of munitions. At the front of the trailer are a retractable hitch and a drawer for storing supplies.

Fig. 3-4   NGMT prototype.

One of the goals of the NGMT was improved ergonomics with respect to usability and maintenance. The height of the trailer bed needed to be optimal for loading and unloading either using forklifts or manually. Recessed tie-down rings were provided to secure the munitions. Placement of these rings had to be both functionally adequate and ergonomically reasonable. Ring placement was considered unacceptable if a person could not reach it while the trailer was fully loaded [7].

AEF provided CAD information in the form of AutoCAD drawing files on a 100-megabyte Zip™ disk. Unfortunately, this information was unusable for several reasons. First, we did not have an AutoCAD-to-Jack translator and did not have funds to purchase one. I searched for a freeware translator, but to no avail. Second, although many CAD files were provided, there did not appear to be a composite file that would be useable for this analysis. Such a file may not have been necessary during the development of NGMT, but was necessary to perform engineering simulations. Even if such a file existed, it probably would have contained too much detail (i.e., surfaces) in some ways and too little detail in others. And even if articulated joints (i.e., moving parts) were modeled in AutoCAD, they probably would not have translated. Therefore a virtual model of the trailer was created using Jack's modeling capabilities. Since the geometry of the trailer was not very complex, and since AEF provided the standard three views for engineering design, I was able to model the trailer, trailer extensions, and munitions (shown in Fig. 3-5) in a fairly short time.

Modeling the moving parts of the trailer was by far the most challenging and frustrating task. The bed of the trailer separated to accommodate large munitions. The front axle and tow bar rotated over 180 degrees in order to steer the trailer. The tow bar also could be retracted when not in use. To articulate the bed's expansion, a translation joint was created. This Jack feature allowed a minimum and maximum translation to be set so that the model could not expand beyond that allowed by the real trailer. It also prevented the two halves from overlapping each other, because in the virtual world, objects can overlap.

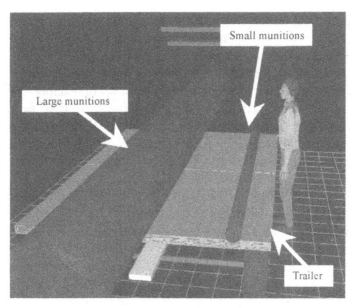

Fig. 3-5 Jack models of trailer, human, and large and small munitions.

Some moving parts and other details were not modeled because they were not important to the analysis. For example, the wheels did not turn but the trailer could still translate (slide) along the ground plane, which was realistic enough. The bed of the trailer seemed to float above the wheels because there was no apparent physical connection; however, using an attachment in Jack, these parts would move as one unit. Although this may have been visually unnatural, these details were not important for the analysis.

Modeling of the munitions was also quite simple. It seemed logical to examine the extreme sizes (i.e., biggest and smallest) and average size rather than all sizes accommodated. The AEF Battlelab told me that the largest munitions had a length of 249 inches, a radius of 25 inches, and a weight of 3200 pounds. The smallest had a length of 113 inches, a radius of 5 inches, and a weight of 195 pounds [7]. Creating cylinders of the proper diameter and length was sufficient for this application, since other details had little effect on the outcomes. Jack provided a simple method to create these cylinders by simply inputting the length and diameter.

3.7 Performing the Analysis

Several aspects of human interaction with the trailer were examined. These included loading munitions onto the trailer, securing the munitions to the trailer bed, extending the trailer, and transporting the trailer itself. Since munitions must be tied down to the bed of the trailer, I evaluated reach distances of a variety of humans with various sizes and placements of munitions (Fig. 3-6). Modeling of the straps and pulleys used for tie-down would have been possible by using the "hose" generator* developed under the DEPTH contract. However, the version of the Jack software being used did not have this capability and updating the software was not feasible due to budget constraints.

Munitions are difficult to work with on any trailer. They can be arranged in various configurations depending on their sizes and the mission requirements. For example, the MHU-12 munitions handling trailer shown in Fig. 3-7 can accommodate a wide variety of munitions using various adapters, cradles and chocks. The trailer is shown with ADU-314/E aerial stores adapters,

---

* Hose generation allows long, flexible entities, such as cables, bundled cables, or fluid hoses, to be created.

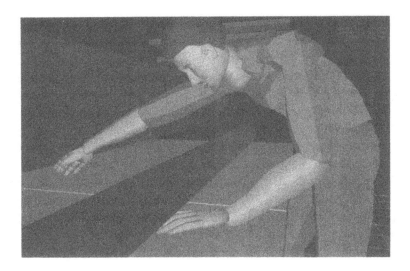

Fig. 3-6 Checking the reach distance of a medium-sized male for placement of munition tie-down rings.

Fig. 3-7 MHU-12 trailer. (Courtesy U.S. Air Force Museum, Wright-Patterson Air Force Base, Ohio.)

enabling carriage of pre-loaded bomb racks rather than individual bombs. The BRU-3 bomb racks are loaded with 500-pound MK82 "Snake-eye" bombs [8].

For purposes of this analysis, I examined how various-sized munitions would be transported on the NGMT. The largest was far too large and heavy for manual handling, so there were no significant manual material handling issues to evaluate. Medium-sized munitions actually presented more interesting human factors challenges. A small female and a medium male were shown to have enough room to stand on the trailer to work around the munition (Fig. 3-8). The only possible problem that could be envisioned was that the female could not reach far enough to tie down the munitions if she attempted to do so from the opposite side of the munitions. The medium male would be able to reach the required location, but the activity might still be unsafe. It was thus recommended that this activity be accomplished from the ground if possible.

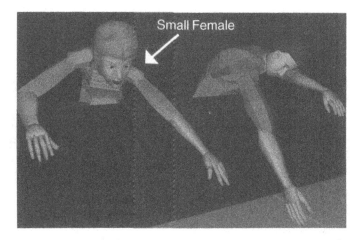

Fig. 3-8 Small female and medium male reaching over medium-sized munitions.

The next step was to evaluate the trailer's ability to expand for large munitions and collapse for small munitions. A couple of simple simulations showed that extending to the trailer bed would not be a problem. The only potential safety concern was the possibility of pinched fingers while collapsing the trailer.

The deployment of the trailer was briefly examined. A simple mockup of a C-17 cargo compartment was constructed and the human model was placed in the most likely working positions as the trailer was loaded. The dimensions of the C-17 compartment were found to be 88 feet long by 18 feet wide by 12 feet 4 inches high, so there was enough room for several trailers [9]. In its most compact state, the trailer posed no human factors problems in loading or unloading onto this compartment.

3.8 Software Observations

The developers of Jack are to be commended for creating such a complex yet useful simulation package. Humans are arguably the most difficult figures to model and simulate, but Jack has done a good job of bringing them to life. However, while Jack was a university product, it was not a particularly reliable software product. We often had to work around problematic features and save work often in case of an unexpected crash. Despite the shortcomings, we found the University of Pennsylvania software to be useful on both the F-15 and F-22 analyses.

EAI's commercial version of Jack used on the NGMT analysis was considerably more robust, plus several useful features were added. The most notable addition was the ability to directly manipulate objects by clicking on the object. Clicking the right mouse button while pointing at an object brought up a menu of applicable actions that could be performed on the object. This may seem like a small convenience, but it turned out to be a real time saver. With previous versions, clicking on an object may in fact have selected another object behind the one intended.

On the downside, EAI's software had a couple of apparent discrepancies which had to be worked around. First, the software segment faulted several times in creating joints on the trailer. This typically occurred while attaching two specific objects, so it could have been that one of these objects was corrupted. Replacing these objects with new objects seemed to fix the problem. Another software discrepancy was that all aspects of the models were not preserved when saving scenes. Some joints that were created in one session, for example, were lost in the next session. This bug added a significant amount of time for rework. These two discrepancies prolonged the task and raised the frustration level. It should be noted that all software programs

51

of this complexity have bugs and, given the amount of functionality in Jack, I was impressed at how stable it actually was.

3.9 The Appeal of Human Modeling

Some people wonder if using human models for maintainability analysis is worthwhile because the modeling and simulation process, in particular the time it takes to acquire the equipment models, can be quite long. Importing models from CAD is somewhat simpler than it used to be, thanks to better translators and some standardization, but this can still be a painful, multi-step process. In some cases, the CAD file will need to be translated into a neutral format before it can be imported into the simulation environment. After the CAD file is imported, it will probably need to be decimated to reduce the geometric complexity; moving joints will probably need to be added.

Working with human models, especially creating full simulations, can also be time-consuming. When you consider that many designs are already ergonomically sound, it can be frustrating to go through all of these steps and turn up nothing. However, the payoff when a problem is discovered can be large, so it is a worthwhile exercise for many equipment designs.

Some argue that it is better to use live humans interacting with full-scale physical mockups than virtual humans interacting with virtual mockups. Physical mockups have been around much longer than computer simulations, and most people have more confidence in live analyses. This lack of confidence in virtual human analysis is not unfounded, because there are many intricacies of humans that are quite difficult to model. One study performed on the DEPTH contract showed that abilities of live humans performing a standing reach varied significantly from that of a virtual human [10]. Sometimes the virtual human indicated a successful reach when the live humans failed, and vice versa. It is suspected that factors such as balance, joint extension, and comfort contributed to the discrepancies. Inaccuracies in the measurement of the live subjects may have also contributed. In fairness, this study focused on extremes in standing reach, and analyses that I have performed do not require such fine accuracy in the reach envelope.

Given all of these drawbacks, why use human models for engineering analysis? The answer is primarily cost and schedule. Physical mockups, even simple wooden replicas, are expensive to fabricate and store. Furthermore, if the design needs to be changed, it is usually costly and time-consuming to modify. Virtual mockups, on the other hand, are relatively fast and easy to acquire and modify. Design concepts can be prototyped and evaluated on the computer early in the design process, long before anything is built. These design concepts can be modified fairly easily and reevaluated.

With a growing trust in computer-based analyses, fewer physical mockups of major weapon systems are being built. In some cases, decision-makers are opting not to build them at all. However, it does not make sense to use human modeling to evaluate every man-machine interaction in complex systems. If a design closely resembles an existing system that is ergonomically sound, human modeling probably is unnecessary. For original designs of equipment with a useful life of five years or more, modeling is probably worthwhile if there is any question about the human-machine interfaces. It is important to point out that the life-cycle cost incurred by poor design can dwarf the cost of the analysis. Also it should be noted that, although ergonomics can be a significant cost driver, few program managers will approve this type of design change in later stages of acquisition when physical mockups are built.

3.10 A Better Virtual World

One can think of digital humans as desktop human factors tools. They allow designs to be analyzed without costly physical mockups or even venturing outside of one's office. Electronic transfer of digital mockups allows geographically distributed teams to analyze designs in ways that were not available a few years ago. Even when a full-scale physical mockup exists, it may not be readily accessible to those who should perform certain evaluations.

So there are clear benefits to human modeling, but there is still room for improvement. The steps to set up the analysis are usually unintuitive and time-consuming [1]. The technology will gain wider use when the steps described below occur seamlessly.

1. **A design file is read in without the user being concerned about geometric detail.** Today, the extraneous data bog down the simulation system, particularly when automatically detecting collisions between

the human and equipment. Files from any 3D graphics system such as CAD or visual simulation should be imported without intermediate conversions.

2. **The user states what task is to be evaluated.** This can be accomplished either with selections from a menu or natural language commands. For maintenance evaluations, this would require the simulation system to understand what it takes to perform tasks such as remove and replace.

3. **Simulations are automatically performed.** The simulations will check a range of humans from small females to large males. The camera (i.e., the user's viewpoint) would be automatically positioned for best viewing. For fine motor tasks, the camera would zoom in.

4. The user could then **refine the simulation** to better suit current needs, again with simple menu selections or natural language.

5. **An easy-to-understand report is provided** with relevant statistics, measurements, and design change recommendations.

6. In place of automatically generating simulations, the user's **movements can be tracked** with virtual reality equipment, allowing sharing of an environment with the virtual mockup. One can think of the user becoming the human model, seeing the avatar's body through a head-mounted display. Virtual objects handled by the human model can be felt through the use of tactile and force feedback devices.

3.11 Conclusions

In this chapter, three analyses of Air Force equipment have been discussed. The F-15 analysis turned up no unexpected human factors problems, but the new functionality for creating task simulations was noteworthy. The F-22 analysis identified some potential problems, especially for very small or large people. The trailer analysis demonstrated that human models could even facilitate the analysis of relatively simple equipment, and that the analysis process was not trivial despite the simplicity of the equipment. The underlying theme in all three analyses was that the technology was not totally mature, but quite useful nonetheless.

Finally, the benefits and shortcomings of using virtual mockups were discussed. Although the analysis process can be difficult and time-consuming, the potential savings in safety, manpower, equipment availability, and reliability make it worthwhile. If a maintenance procedure can be streamlined by even a few minutes, significant savings will result when the procedure is performed thousands of times over the system's life.

References

1. McDaniel, Joseph W., "Human Modeling: Yesterday, Today, and Tomorrow," SAE International Conference, Digital Human Modeling for Design and Engineering, Dayton, Ohio, 1998.
2. Lane, Ken and Ianni, John D., "Design Evaluation for Personnel, Training and Human Factors (DEPTH) Final Report," AFRL/HE-WP-TR-1998-0007, 1998.
3. Engineering Animation, Inc., *Jack Human Modeling and Ergonomic Analysis System, User's Guide,* 1998.
4. Vujosevic, Ranko and Ianni, John D., *A Taxonomy of Motion Models for Simulation of Maintenance Tasks,* AL/HR-TP-1996-0045, 1996.
5. Karger, Delmar W., *Engineered Work Measurement,* Industrial Press, 1977.
6. Badler, N.I.; Phillips, C.; and Webber, B., *Simulating Humans: Computer Graphics, Animation, and Control,* Oxford University Press, New York, 1993.
7. "Air University, Munitions Systems Journeyman, Volume 2. General Handling and Support Equipment," CDC 2W051B, date not available.
8. McDonnell Douglas Corporation, *Technical Manual: Loading and Air Transport of Nuclear Cargo (Nonpalletized),* 1996.
9. U.S. Air Force, *C-17 Globemaster III Fact Sheet,* Air Mobility Command, Public Affairs Office; 502 J Street; Scott AFB, Ill.
10. Nemeth, K.; Ianni, John; and Wampler, Jeff, "Evaluation of Human Model Standing Reach," AFRL/HE-WP-TR-1999-0227, 1999.

Chapter 4

Ship Bridge Design and Evaluation Using Human Modeling Systems and Virtual Environments

Patrick A.J. Punte, Aernout J.K. Oudenhuijzen
and Andrea J.S. Hin
TNO Human Factors
The Netherlands

4.1 Introduction

For many years, TNO Human Factors has been involved in the process of designing ship bridges and other operational units for merchant vessels as well as frigates and other vessels for the Royal Netherlands Navy. During recent years there has been an increased use and acceptance of automation aboard ships. Various functions such as machinery monitoring and control, cargo monitoring, communication, and emergency handling, have increasingly been concentrated on the ship bridge. The initial motivation for these changes often has been the reduction of crewmembers in order to reduce operational costs. As a result, information displays and controls must be integrated in a "relatively small" workplace to accommodate the few crewmembers present.

Recent technological developments enable the integration of separate ship systems, for example, the integration of electronic chart display and information systems (ECDIS) and radar. The integration of such systems does have an impact on the layout of ship bridges. In general, a higher level of automation and integration opens the way to a more compact bridge design. A compact design is very important to guarantee the performance and safety of the complete system with small crew sizes. It is evident that the increasing automation and concentration of functions on ship bridges requires a thorough study to achieve an optimal design.

4.2 Ship Bridge Design

Human factors engineering has to be an integrated part of the total design process to obtain an optimized design. The human factors engineering process starts with an analysis of the mission of the product or workplace to be designed. For example, the mission of the ATOMOS II chemical tanker was "fast, safe and economical transportation" [1]. Based on this mission, a function analysis has been performed (Fig. 4-1). The purpose of the chemical tanker was divided into six main functions: "travel," "communicate," "maintain platform," "monitor and maintain cargo status," "monitor passenger and crew status," and "anticipate emergency." The function analysis has a hierarchical structure. Functions are broken down until a level is reached at which the specified functions are assigned to a human or an instrument (Fig. 4-1). At this level, the functions are called tasks and the process of assignment is called task allocation. An example of such a task is "gather meteo information" (Fig. 4-1). Task allocation studies result in a specification of personnel and instruments that are used as input for the design of the workspace.

After the mission analysis, function analysis, and task allocation, the actual geometric design can be made. During the design of ship bridges, the following ergonomic aspects are relevant [2]:

- General layout of the bridge:
  Aspects belonging to this category are the spatial arrangements, maintenance, safety and accessibility of the bridge, and inside and outside view at the bridge.

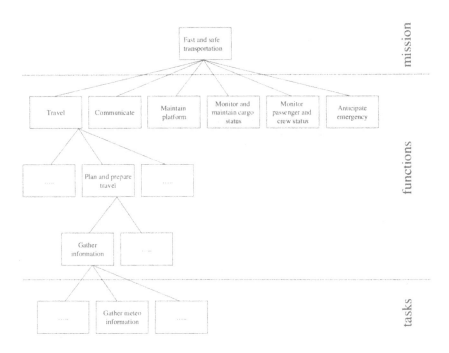

*Fig. 4-1 A part of the function of the ATOMOS II chemical tanker.
One of the main functions is "travel." "Plan and prepare travel" is
a sub-function of "travel," etc. An example of the lowest level is
"gather meteo information," which is defined as a task.*

- Workstations:
  Aspects belonging to this category are reachability, visibility, and read-ability of instruments, maintenance and physical workload at the workstations.

- Work environment:
  Aspects belonging to this category consider the ship motion, lighting, sound, noise, climate, and vibrations.

- Man-machine interface:
  Aspects belonging to this category are functionality, usability and organization of interfaces of instruments, and displays and mental workload as a result of these interfaces.

59

- Team aspects:
  This concerns the cooperation of various operators at the ship bridge, for instance during extreme scenarios, such as man-overboard or replenishment at sea.

While integrating these aspects with specifications put forward by constructors and suppliers, usually compromises must be found. As a consequence, the design most often cannot be optimal in all its aspects. The design, therefore, should be thoroughly evaluated.

4.3 Ergonomic Evaluation of the Design

Various techniques are available to support an ergonomic evaluation of a design. A traditional technique for evaluating a ship bridge design is to make use of a full-size wooden mockup (Fig. 4-2). In the mockup, today's and

Fig. 4-2 Officers of the Royal Netherlands Navy evaluating the mockup of the ship bridge of the Air Defense and Command Frigate. The console at the right is the navigation console which is located at the front of the bridge. This evaluation focused on the layout of the bridge and the design of the consoles. The officers had to pass through imaginary scenarios systematically, such as mooring or replenishment at sea, and note their comments on an evaluation form.

future operators perform tasks according to various scenarios. This procedure stimulates discussion and evaluation by designers and operators. As a result, observations during the evaluation may lead to adjustments of the design. It is obvious that the application of a mockup is very labor-intensive, time-consuming and, as a consequence, offers little flexibility regarding modifications.

Advances in computer technology, especially computer aided design (CAD) and visualization techniques, have made it possible to replace wooden mockups partially by computer-assisted ergonomic analysis methods. These analysis methods enable human factors engineers to verify the ergonomics of workspace designs very early in the design process. The computer-assisted ergonomic analysis at TNO Human Factors is based on the use of two means: virtual environments (VE) using head mounted displays (HMD) and human modeling systems (HMS).

The application of virtual environments offers the possibility of evaluating the outside and inside view at a ship bridge. Outside view is an important issue during ship bridge design to guarantee the primary goal: safety. An evaluation using a virtual environment stimulates participation of (future) operators, customers, and designers in an early stage of the design process. As in mockups, the performance of operators can be tested in various scenarios.

Haptic feedback is lacking in a virtual environment. For instance, in a VE one can easily walk through walls. To overcome this problem a simple wooden mockup, e.g. the boarding of a bridge, can be combined with a VE that is exactly lined up [3]. This technique, called the hybrid mockup, was successfully used for the design of the bridge of the Air Defense and Command Frigate of the Royal Netherlands Navy, which is expected to be in operation in the year 2001 (Fig. 4-3).

As compared to a VE, an HMS is more suitable to evaluate the anthropometric aspects of a design, for instance the accessibility of rooms or reachablity of instruments. As part of the project "Human Modeling Systems in Ship Design," TNO Human Factors investigated the possibilities for the application of HMSs at the design department of the Royal Netherlands Navy [2]. The value of HMSs has been compared with ergonomic information coming from existing tools that are used during the design process of ships

Fig. 4-3 Officer of the Royal Netherlands Navy in a hybrid mockup. The officer is standing at the bridge wing, looking at the quay during an onmooring procedure.

(design drawings and models, handbooks and norms, mockups, virtual environments, and hybrid mockups). For each phase in the design processes of ships, the most useful tools were selected. It has been concluded that HMSs have an added value in the design process of ships, especially during the design of room-layouts and workstations. To evaluate outside views from a ship bridge, virtual environments or hybrid mockups are preferred. To evaluate a design with several operators playing their role in a scenario, the application of full-size wooden mockups is still the best way to go.

4.4 Computer-Assisted Ergonomic Analysis: The ATOMOS II Project

Ergonomic aspects of a future bridge of a chemical tanker were analyzed using computer-assisted ergonomic analysis (CAEA). This ship bridge was designed as a part of a European project (ATOMOS II). The main question to be answered in this project was which conceptual layout of the ship bridge would enable the crew to perform their tasks in operating the ship in a safe, efficient, and comfortable manner. The answer to this question has been given in terms of physical geometric aspects and functional aspects of the design.

The CAEA of the ship bridge was divided into two parts:

- The physical analysis concerned the dimensions of the bridge in relation to the anthropometry of crewmembers. This anthropometric study has been carried out using the Boeing Human Modeling System (BHMS) and the CAD tool ProEngineer [4].

- The functional analysis concerned the functional relations between crewmembers at the ship bridge and the outside and inside view. Evaluation of these functional aspects has been supported with VE [5].

4.5 Physical Analysis Using BHMS

4.5.1 Bridge Design
Fig. 4-4 shows the first conceptual design of the ship bridge.

4.5.2 Method

The CAD model of the ship bridge was translated to BHMS using conversion to the format of stereolithography (STL). To evaluate the design with BHMS, a crew population was defined based on appropriate anthropometric databases on Europeans. The crew population is subject to changes. Among those are the trends of secularization and an increasing occupation of ship bridges by women. This results in a large and growing range of anthropometric data which must be taken into account in the design of workspaces. The lower design limit is dictated by anthropometric data of small women in the

Fig. 4-4 First conceptual design of the bridge. This design was made using the CAD software ProEngineer. A navigation console is positioned at the front of the ship bridge. This console consists of displays for radar, ship's own data, electronic charts, and machinery control systems.

year 1997. The upper design limit is dictated by the tall males in the year 2015. The reason for using estimated anthropometric data for the year 2015 was to anticipate the life-cycle time of ships.

The workstations at the ship bridge were designed to enable standing as well as sitting postures. To evaluate the design in standing and sitting postures, two critical dimensions of a human are selected: stature and sitting height. These parameters are varied to compose a population of twelve manikins in BHMS. Statures of these manikins vary from 1551 mm (smallest female in 1997) up to 2026 mm (tallest male in 2015), Table 4-1.

4.5.3 Results

All manikins were positioned at the navigation console of the ship bridge. The following aspects were evaluated:

- sitting and standing postures;
- reachability and visibility of displays and controls;
- outside view;

Furthermore, requirements for adjustment ranges of the chairs and accessibility of the bridge were imposed. Fig. 4-5 shows manikins standing or sitting behind the navigation console of the ship bridge.

Table 4-1 Visibility and reachability of manikins standing behind the navigation console of the ship bridge. A '+' sign means no problems are expected.

Manikin	Stature (mm)	Sitting height (mm)	Visibility: Downward gaze direction (degrees)	Reachability	
				Keyboard	Phone
M1 (female)	1551	830	5°	too high	bend trunk 10° forward
M2 (female)	1580	815	7°	+	bend trunk 10° forward
M3 (female)	1580	868	7°	+	bend trunk 11° forward
M4 (female)	1683	837	15°	+	+
M5 (female)	1683	923	15°	+	+
M6 (male)	1785	875	21°	too low	+
M7 (male)	1785	977	21°	too low	+
M8 (male)	1887	915	27°	too low	+
M9 (male)	1887	1025	27°	too low	+
M10 (male)	1990	977	33°	too low	+
M11 (male)	1990	1052	33°	too low	+
M12 (male)	2026	1029	34°	too low	+

Table 4-1 summarizes some of the results. The first three columns indicate the body dimensions of the crew population. As already stated, the two driving body dimensions were stature and sitting height. The fourth column shows for each manikin the downward gaze direction to the rim of the console. The criterion was a downward visibility of 12 degrees or more under the horizon. During the generation of the design requirements, and the realization of the first conceptual layout, not enough attention was paid to the obstruction of view by the rim of the console. The rim appeared to be too high for an unobstructed view by the three smallest female operators. The design, therefore, had to be adjusted.

The fifth and sixth columns of Table 4-1 show the results regarding reachability of keyboard and phone at the navigation console. Comments on reachability of

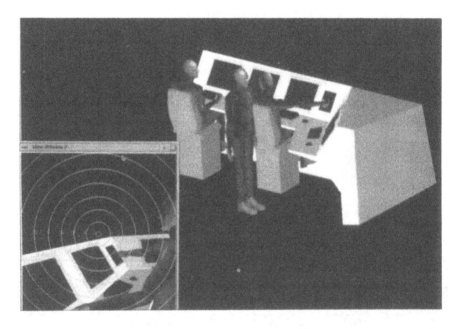

Fig. 4-5 Manikins sitting and standing behind the navigation console. The window in the lower left-hand corner shows the view of the sitting manikin at the left (BHMS). This console is suited for standing as well as (heightened) sitting postures. This console is the same as the console in Fig. 4-4.

the keyboard derived from the height of the elbow and length of the lower arm. The judgement of keyboard reachability is based on the 10-fingers typing system. The keyboard was positioned too high for the smallest female manikin, and too low for the male manikins. Another criterion was to reach the phone without bending the trunk. The results show that for the three smallest female manikins a forward bend of the trunk was required to reach the phone properly. After this evaluation, the instruments were repositioned to obtain a design in which the criteria were valid for the whole crew population.

Another problem that appeared from this analysis was the height of door openings. In the first design, the height of door openings was based on the statures of the current target population. However, the physical analysis made it clear that the first design did not satisfy an unobstructed accessibility for taller operators in the year 2015.

4.5.4 Discussion

Most of the problems that appeared from the physical analysis were the result of the large range of body dimensions of the expected user population, which included women and users in the year 2015. Subjects with extreme body dimensions are hard to recruit for mockup evaluations. Using HMSs, it is even possible to evaluate the design using a future target population. One of the disadvantages of HMSs is the lack of feedback when a manikin is positioned in an uncomfortable way. The designer has to recognize these unfavorable postures by himself.

It is concluded that a physical analysis can be carried out easily during the design process and alterations are implemented quickly due to an easy conversion between the CAD tool and the HMS.

4.6 Functional Analysis Using VE

4.6.1 Bridge Design

Based on the results of the physical analysis, the ship bridge was modified. The improved design of the bridge has been used as input for the functional analysis.

4.6.2 Method

Virtual environment techniques were used to carry out the functional analysis. A virtual environment system enables an operator to interact with a three-dimensional computer-generated environment. In current systems, computer graphics are presented through head-mounted displays positioned just in front of the eyes. Head movements are recorded continuously in order to present the correct image from the measured point of view with a fast update rate.

To evaluate the functional aspects of the ship bridge, a chemical tanker was modeled. Furthermore, the VE includes a marine environment for a realistic outside view (Fig. 4-6).

Five operators participated in the VE trials. Each of the participants has or had active duty on a ship bridge. They were immersed for about 45 minutes to evaluate the design. For this, they were positioned on various spots to

Fig. 4-6 Two images of the virtual environment. The ship, a chemical tanker, is located at the harbor. Left: outside view sitting behind the navigation console. From this position, a good view at the bow of the ship is possible. Right: view inside the bridge, standing at the right bridge wing.

udge the design. The VE trials also included a "walk through" steered by the participant using a so-called flying joystick. A dynamic scenario was introduced to evaluate external visibility during unmooring, passage at open sea, and mooring.

During the VE evaluation, feedback was obtained by open discussion, thinking aloud. All issues addressed were noted down. During a final discussion, these issues were recalled. The participants discussed until they agreed on a solution or found a compromise.

4.6.3 Results

During the VE trials and the discussion, a large variety of issues were brought up.

One of the main issues during the VE trials was the visibility from the ship bridge. As a result of the decreasing number of operators on duty, a good outside view should be guaranteed from all positions at the bridge. Making use of VE, participants were able to evaluate the outside view from various spots, for example, standing and sitting behind the navigation console and standing at the bridge wing. As a result of the trials and the discussion, an improved, more compact shaped bridge was designed. Enlarged windows (floor to ceiling) were introduced to improve the outside view.

Another main point of discussion concerned the bridge wings. Originally, the ship bridge was presented with open bridge wings. During mooring and unmooring, operators at the bridge wing felt a strong need for information on the velocity of the ship and its direction of movement. The advantage of an open bridge wing is the unrestricted outside view. In particular, the possibility of leaning over the boarding improves the outside view. This is not possible on a closed bridge wing as a result of the windows and window posts. However, in unfavorable weather conditions, such as tropical heat and stormy weather, closed bridge wings are preferred. When the need for information would be fulfilled by the application of additional displays at the wing, and when the windows would be repositioned to optimize the outside view, a closed bridge wing is a good alternative. It was decided to redesign the ship bridge with closed bridge wings, including bridge wing control panels to support maneuvering at the bridge wing.

During the trials, a discussion came up about the displays and controls at the navigation console. During the discussion, a final set of instruments and displays was established.

Based on the outcome of the VE trails, the final design of the future ship bridge of a chemical tanker was created (Figs. 4-7 and 4-8).

Fig. 4-7 Final design of the bridge mounted on top of the accommodation tower of the chemical tanker.

4.6.4 Discussion

The results from the VE trials depend largely upon the skills and experience of the participants. Participants were involved in the discussion of, for instance, open or closed bridge wings. Also, creative solutions were generated on the fly, for instance to equip the bridge with a height-adjustable floor to accommodate each individual working on the bridge.

From the remarks on the designs made during the trials and the discussion afterwards, VE is to be regarded as a valuable tool in the design process. It

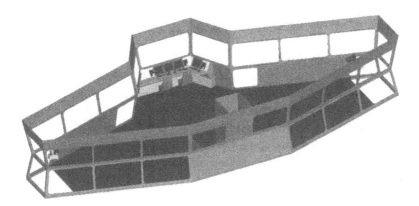

Fig. 4-8 Final design of the bridge. As compared with Fig. 4-4, the design of the bridge has changed. The shape of the bridge has been optimized for a good view sitting behind the navigation console. Both bridge wings have been added to the bridge itself. The geometry of the windows has been modified to guarantee an optimal view at the quay during, for instance, mooring and unmooring.

supports participation of (future) operators, customers, and designers in an early stage of a design process. Nevertheless, haptic feedback is lacking in the current VE. For instance, in a VE one can easily walk through walls. To overcome this problem, a hybrid mockup can be used [3].

4.7 Discussion and Conclusions

This section describes the advantages and disadvantages of the application of the CAEA as compared with traditional design and evaluation (for instance handbooks or wooden mockups).

Wide Range of Anthropometrics

Using an HMS, it is possible to take into account the crew population, including subjects to be expected as future crew members. Live subjects of all anthropometric extremes are very hard to find for traditional mockup evaluations. Another advantage of an HMS is the safety of subjects. Using an HMS it is possible to evaluate poor designs without possibly harming subjects.

Participation of Operators and Customers

Visualization of a design stimulates the exchange of information very early in the design process, when the first concepts are made. Using these concepts, CAEA methods stimulate the participation of designers, (future) operators, and customers in the design process. This makes it possible to verify the ergonomic qualities of the design using the skills and experiences of operators. VE techniques make it possible to situate a design in a realistic setting, which enables evaluation of aspects that cannot be evaluated using wooden mockups.

Risk Reduction

Using CAEA there will be a reduction of risk during the design process. Design errors causing serious ergonomic problems quickly become visible from the application of HMSs and VEs. For example, during the VE evaluation of the Air Defense and Command Frigate of the Royal Netherlands Navy, viewing problems caused by a radar antenna emerged quickly [3]. These problems were not directly perceived by CAD models alone. Therefore, the application of VE prevented costly design modifications after the ship has been buit.

Reduction of Time and Costs

In comparison with a wooden mockup, modification of an electronic model of the design following an evaluation can be carried out more quickly and cheaply. For instance, a traditional evaluation of a ship bridge using a wooden mockup costs approximately 50,000 US dollars and takes about 4 months to build and evaluate. The electronic evaluation, including the application of a human modeling systems and a virtual environment, costs 20,000 U.S. dollars and can be carried out in 1 month. In this case, it is even possible to change the design during the evaluation. CAEA offers more flexibility during evaluations and is much quicker as compared to the traditional design methods.

Of course, there are also some disadvantages of CAEA as compared with wooden mockups.

Haptic Feedback

A wooden mockup provides haptic feedback, which enables an operator to evaluate, for instance, reachability of instrument panels or accessibility of consoles. VEs do not provide haptic feedback, such as interference with objects, because consoles and walls only exist digitally or virtually. The introduction of a hybrid mockup, a combination of a VE and a simple wooden mockup, partly overcomes these problems [3].

Team Evaluations

Wooden mockups are to be preferred when complex scenarios have to be evaluated with more than one operator playing a role simultaneously, especially for evaluating communication by face-to-face contact.

Effects of Ship Movement

Current human modeling systems cannot be used to evaluate the effect of ship movement on the work performance of the operators on the ship, for example, the effect of heavy ship movements as a result of the weather on the equilibrium of an operator while walking on the bridge. To carry out these kinds of studies, TNO Human Factors uses a ship movement simulator (SMS).

Validation of HMSs and VEs

At this moment, HMSs are not validated and verified to their finest detail. However, current HMSs are already useful first order representations of the human body for ergonomic workspace design. Traditional anthropometric databases consist of tables with dimensions such as stature, sitting height, knee height sitting, etc. Often other dimensions are needed for workspace design: dimensions in 3D space or combined anthropometric dimensions. Most of these dimensions are not provided by the traditional databases. HMSs provide these needed dimensions in the form of a 3D manikin.

It is concluded that the application of HMSs and VEs during workspace design will result in a quicker design process as well as in a better design. The techniques currently available partly replace expensive and elaborate building of full-scale mockups.

4.8 Future Work

Future work on computer-assisted ergonomic analysis can be divided into the following parts:

- The ongoing investigation of the application of human modeling software as part of the ship design process will be continued. Various HMSs will be compared using typical case studies for ship design. Using these case studies, a method will be developed which describes how to use an HMS during different stages in the design process of ships.

- TNO Human Factors and the U.S. Air Force combined efforts in the verification and validation of HMSs. The aim is to test commercially available HMSs. The F-16 crew station is used as the application area for validation and verification.

- The full integration of CAD, HMS, and VE functionality will improve flexibility of applying CAEA in evaluating designs. Several commercially available CAD software packages already have some kind of a human model. However, functionality of the human model is often subordinate to the design functionality. Developers of HMSs have initiated new developments in creating the possibility of including their human model and functionality as a module in CAD software. An example is the HMS software RAMSIS, which can be integrated in the CAD software CATIA and ICEM-DDN. A similar integration is developed for VEs and HMSs. Here, the integration of the HMS software SAFEWORK with the VE software dVISE can be mentioned as an example. In virtual environments, the user is always playing the key role. His virtual body is often defined partly, e.g., only the hands, despite the fact that visual feedback on movement of his own body enhances the sense of presence. Completeness of the virtual body is considered an important feature.

- A study on natural human motor behavior is being carried out. The project's focus is on static gazing and reaching during sitting and standing. Results will be presented in algorithms that are ready for implementation in HMSs.

- In cooperation with TNO Automotive, a comfort prediction model will be developed to assess the effects of vehicle dynamics early in the design.

References

1. van Doorne, H. and Schuffel, H., "Function and Task Analysis," ATOMOS II, Task 1.2.1, ID Code: A212.01.10.052.002, TNO Human Factors Research Institute, Soesterberg, 1996.
2. Punte, P.A.J. and Hin, A.J.S., "Human Modeling Systems in Ship Design, Phase 1—Ship Design and Human Factors," Report TM-99-A079, TNO Human Factors Research Institute (in Dutch), 1999.
3. Werkhoven, P.J.; Post, W.M.; and Punte, P.A.J. "Validation of ADCF Bridge Concepts Using Virtual Environment Techniques," Proceedings of the Eleventh Ship Control Systems Symposium, Vol. 1, Department of Ship Science, University of Southampton, Southampton, 1997.
4. Punte, P.A.J.; Oudenhuijzen, A.J.K.; and Hin, A.J.S., "Design of the Layout of Standardized Ship Control Centers," ATOMOS II, Task 1.2.3-4, ID Code: A212.03.10.055.006, TNO Human Factors Research Institute, Soesterberg, 1998.
5. Hin, A.J.S.; Punte, P.A.J.; Bakker, N.H.; and Oudenhuijzen, A.J.K., "Virtual Environment Evaluation of Ship Control Center Concepts," ATOMOS II, Task 1.2.3-3, ID Code: A212.03.10.052.005, TNO Human Factors Research Institute, Soesterberg, 1998.

Chapter 5

Using Digital Human Modeling in a Virtual Heavy Vehicle Development Environment

Darrell Bowman
International Truck and Engine Corporation
Fort Wayne, Indiana

5.1 Introduction

With the availability of better-defined anthropomorphic models, the use of digital human simulations in product development for the military and industry is growing at a staggering pace. These digital human simulations are being employed as a tool to bridge the gap between the virtual model-based development environment and the reality of the end-user. The benefits of using virtual human modeling in heavy commercial vehicle development are twofold. First, human simulation reduces the iterative cycle time between concept development and functional validation. The design characteristics of a system can be analyzed for functional compatibility in a quicker, less expensive, and reliable manner. Second, human simulation increases the heavy vehicle manufacturer's competitive advantage through more innovative designs and improved user acceptance, safety, and usability.

Due to the National Traffic and Motor Vehicle Safety Act of 1966, vehicle and vehicle component manufacturers are required to certify the safety of every vehicle manufactured. The two primary areas of heavy commercial vehicle product development to which digital human models can be applied include proper packaging of the driver's environment and vehicle cab entry and egress. Both of these areas require verification documentation of compliance using a sample of the intended customer population for both the federally mandated motor vehicle certification process and for product liability purposes.

The essential requirements for the first area, proper packaging of the driver's environment, is found in the Federal Motor Vehicle Safety Standard (FMVSS) 101, Controls and Displays [1]. These requirements include criteria for location, identification, and illumination of motor vehicle controls and displays. Digital human modeling can be used early in the design process to verify and document compliance with location requirements using reach and visibility tools, saving both time and costs related to vehicle safety standard compliance.

The second area, vehicle cab entry and egress, is one of the leading contributors to the trucking industry's worker's compensation injuries. To ensure driver and passenger safety, vehicle manufacturers have design requirements to reduce the likelihood of slips and falls from the vehicle during entry/egress. These requirements are based on the guidelines set forth by the Federal Motor Carrier Safety Regulation (FMCSR) Part 399 Subpart L, Step, Handhold, and Deck Requirements for Commercial Motor Vehicle [2] and the American Trucking Association's Maintenance Council Recommended Practice 404B [3]. Again, the reach, joint range of motion, and center-of-mass analysis of digital human modeling are used early in the design process to verify and document compliance with these design requirements.

The text that follows examines the feasibility of applying digital human modeling to the vehicle safety certification early in design, discusses briefly the advantages and disadvantages of this technology in a product development environment, and reviews the foreseeable trends in the industry.

5.2 Federal Motor Vehicle Safety Standard (FMVSS) Compliance

The purchaser of a motor vehicle is entitled to the assurance that the vehicle is safe to operate and that it meets federal safety standards. In 1966, Congress enacted the National Traffic and Motor Vehicle Safety Act to provide the customer this type of protection. This Act requires all vehicles and vehicle component manufacturers to certify that each vehicle produced conforms to all applicable Federal Motor Vehicle Safety Standards (FMVSS) [1].

There are thirty-one Federal Motor Vehicle Safety Standards contained in Section 49 of the Code of Federal Regulations under Part 571. The first standard, FMVSS 101 Controls and Displays, is composed of the location, identification, and illumination requirements for motor vehicle controls and displays. The intent of this standard is to minimize the safety hazards associated with the diversion of the driver's attention or incorrectly selected controls. The requirements of this standard were developed to ensure the accessibility and visibility of motor vehicle controls and displays and to facilitate their actuation under all vehicle lighting conditions. This standard applies to a broad range of vehicles which includes passenger cars, multipurpose passenger vehicles, trucks, and buses. This standard simply requires that the controls listed in Table 5-1 must be operable by the driver who is restrained by the crash protection equipment provided by the vehicle manufacturer [1].

In the case of displays, this standard states that the driver while properly restrained by the vehicle's crash protection equipment, must be able to see all the displays listed in Table 5-2 [1].

To comply with these regulations, each vehicle manufacturer must develop and maintain technical information and documentation related to the performance and safety of each vehicle offered for sale. This technical information is compiled from numerous sources, depending on the focus of the standard, such as engineering studies, installation drawings, laboratory test reports and detail drawings. With the requirement that all motor vehicle configurations must comply with FMVSS, engineering studies are conducted to determine that the vehicle configurations to be tested that will encompass the full range of manufactured vehicles. Once these vehicle configurations are identified, then compliance is documented through individuals demonstrating vehicle usage during laboratory testing. The individuals chosen for this demonstration must represent the anthropometric variability within the intended design population. For those vehicle configurations not selected in the engineering studies for demonstrations, installation drawing are used to confirm compliance through

Table 5-1 Applicable controls

Control Type	Control Description
Hand-operated	Steering wheel
	Horn
	Ignition
	Headlamp
	Taillamp
	Turn signal
	Illumination intensity
	Windshield wiper
	Windshield washer
	Manual transmission shift lever
	Windshield defrosting and defogging system
	Rear window defrosting and defogging system
	Manual choke
	Driver's sun visor
	Automatic vehicle speed system
	Highbeam
	Hazard warning signal
	Clearance lamps
	Hand throttle
	Identification lamps
	Windshield washer
	Windshield wiper
Foot-operated	Service brake
	Accelerator
	Clutch
	Highbeam

Table 5-2 Applicable displays

Speedometer
Turn signal
Gear position
Brake failure warning
Fuel
Engine coolant temperature
Oil
Highbeam indicator
Electrical charge

geometric and mathematical computations. Finally, detailed drawings of the system level components and labeling are used to determine compliance with the identification portion of the safety standard.

Although these regulations provide assurance for customers when purchasing a motor vehicle, there are several negative implications for the vehicle manufacturers. Before the use of virtual reality and digital humans, an actual physical vehicle had to be fabricated before any regulatory compliance could be confirmed. In addition, numerous physical vehicles with a variety of vehicle configurations had to be fabricated to determine compliance across all possible vehicles to be manufactured. Finally, an assortment of people with varying anthropometric dimensions had to be recruited and paid to demonstrate compliance. The following section will present a method for demonstrating FMVSS compliance through a virtual vehicle environment using digital humans which will help alleviate the drawbacks associated with physical testing.

5.3 FMVSS Compliance Analysis

With digital human modeling software, the reach and vision requirements of the Federal Motor Vehicle Safety Standards can be evaluated. Currently, International Truck and Engine Company is using Engineering Animation, Inc. (EAI) Jack® to conduct vehicle product analyses. The crucial element and key to competitive advantage in product analyses using any digital human simulation is the development of the manikins. Therefore, many of the details essential to their creation cannot be revealed in this medium. However, all the manikins applied in the analyses of this chapter are based on the multivariate approach using boundary conditions of key anthropometric dimensions and not the traditional anthropometric percentiles (i.e., 5th, 50th, and 95th). Additionally, the vehicle used in each of the following analyses was specifically created for the purposes of this chapter and does not exist outside the virtual environment.

5.4 Instrument Panel Reach Analysis

According to the location requirement of FMVSS 101, all hand and foot controls listed in Table 5-1 must be operable by the driver while restrained by the crash protection equipment supplied by the vehicle manufacturer [1]. To demonstrate regulatory compliance, a series of virtual human models were postured in a vehicle with the heel and hips constrained to the accelerator-heel point and the appropriate hip point, respectively.

Once these manikins are properly positioned within the vehicle, the reach to each of the identified controls can be evaluated using either generic reach envelopes or a target-specific reach path. The generic reach envelope provides a quick method for evaluating the reach capability to numerous controls simultaneously. Yet, reach envelopes generated in the human modeling software do not account for variations in human reach behaviors due to psychomotor behaviors or obstructions in the reach path. Therefore, reach envelopes are adequate for evaluating a group of controls that only require a straight-line reach without reach trajectory deviations due to obstructions.

For instances when vehicle obstructions will be encountered during the reach, the reach capability should be evaluated by manipulating the manikin's limb manually through a specific reach trajectory. Although this will not account for individual reach trajectory variation, this technique allows the practitioner to adjust the manikin's reach path by navigating around obstacles and providing a more realistic estimation of the manikin's reach capability.

Ideally, the digital human modeling software should provide this target-specific reach functionality automatically given the software's ability to provide collision detection between the humanoid form and vehicle components. The software user would simply select the desired site to be the termination of the reach trajectory, and then the software would solve for the most optimum path given biomechanical constraints and collision avoidance. Although this process would provide a reach path optimized for geometrical constraints and not for human behavior variation, this function can be replaced with reach prediction equations once empirical research of human reach behaviors has been completed.

For the purposes of summarizing this vehicle's evaluation, the more important controls from Table 5-1 were chosen to demonstrate both hand and foot reach paths for different region of the driver's environments. These chosen controls included the windshield defrosting and defogging system (Figs. 5-1 and 5-2), vehicle ignition (Figs. 5-3 and 5-4), driver's sun visor (Figs. 5-5 and 5-6), and clutch pedal (Fig. 5-7). The reach analyses of these controls indicated that there were no controls outside normal reach envelopes for any of the manikins used in this evaluation. However, the reach analysis to the windshield defrosting and defogging system, as seen in Fig. 5-2, revealed an interference between the individual's arm and the manual transmission shift selector.

Fig. 5-1 Generic reach envelope for windshield defrosting and defogging system.

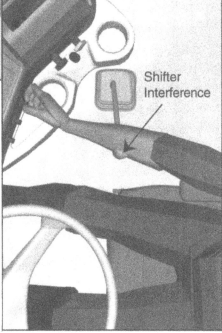

Fig. 5-2 Target-specific reach path for windshield defrosting and defogging system.

Fig. 5-3 Generic reach envelope for vehicle ignition control.

Fig. 5-4 Target-specific reach path for vehicle ignition control.

Fig. 5-5 Generic reach envelope for driver's sun visor.

Fig. 5-6 Target-specific reach path for driver's sun visor.

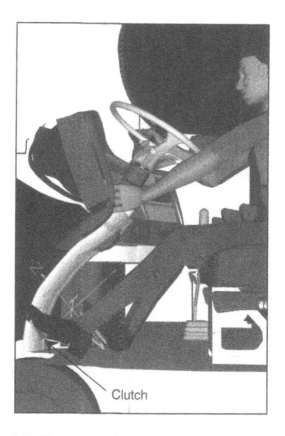

Fig. 5-7   Target-specific reach path for clutch pedal.

5.5 Instrument Panel Visual Analysis

According to the location requirement of FMVSS 101, all displays listed in Table 5-2 must be visible by the driver while restrained by the crash protection equipment supplied by the vehicle manufacturer [1]. As with the reach analyses, a series of virtual human models were postured in the vehicle with the heel and hips constrained to the accelerator-heel point and the appropriate hip point, respectively, to confirm compliance with the vehicle safety standard. Once the manikins are properly positioned within the vehicle's driver's environment, the human modeling software's vision analysis tools can be used to evaluate available display visibility.

The vision analysis tool available in EAI Jack® software is very straightforward to use. The view from the manikin's eyes can be easily displayed and captured by selecting the Eye View option under the Human menu heading. Although this simple visual analysis tool does not incorporate all the characteristics of the eye, it does provide the different views of the manikin's eyes, including between eyes, left eye, right eye, both eyes, or eye view cones.

For the purposes of summarizing the results of these visual analyses, the cluster of primary displays will be the focus. As seen in Figs. 5-8 and 5-9, the visual analyses of this vehicle indicated that all the displays listed in Table 5-2 were easily viewable by both the driver's left and right eyes with minimal head motion. In addition, the analyses (Fig. 5-10) proved that these primary displays were within the manikin's 30-degree field of view.

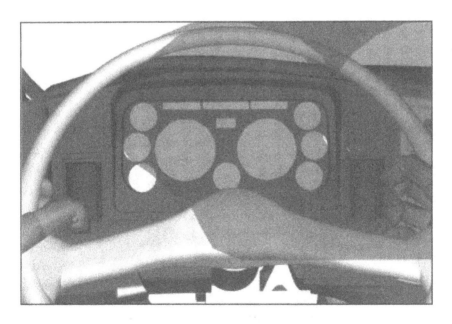

Fig. 5-8 Left eye view of primary display cluster.

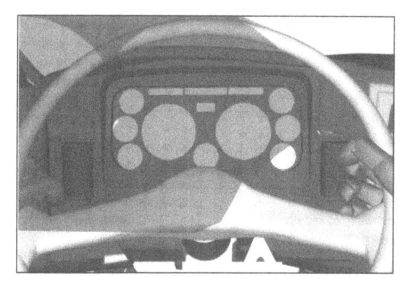

Fig. 5-9 Right eye view of primary display cluster.

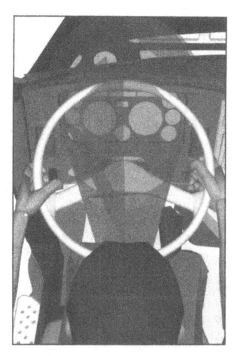

Fig. 5-10 Top view of left and right eye 30 degree view cones.

5.6 Vehicle Entry/Egress

For decades, the task of entering and exiting heavy vehicle cabs has produced physical injuries for drivers, workers' compensation claims for truck fleets, and product liability for vehicle manufacturers. Based on 1997 workers' compensation claims greater than $1,000.00, slips and falls accounted for 34% of all trucking industry "on-the-job" injuries (Fig. 5-11). Of these slip and fall injuries, one-third were attributed to entry/egress of vehicle cabs (Fig. 5-12) [7]. Therefore, vehicle manufacturers spend vast resources designing and manufacturing vehicle steps and handholds before the vehicle is released to the customer. The scrutiny given to the entry/egress system includes a physical demonstration of step and handhold usage which must be completed before the vehicle is released for production. With the advent of digital human modeling, this demonstration can occur earlier in the design phase, before physical representations of the product are developed. Furthermore, the compliance demonstration can include more vehicle configurations than those identified by using either a worst case vehicle configuration or the vehicle with the forecasted highest sales volume.

To reduce the potential for injury, digital human modeling can be used to ensure the physical accommodation and stability of the individual as he/she ascends into or descends out of the vehicle cab. The key ergonomic objective of physical accommodation is to provide adequate reach to both steps

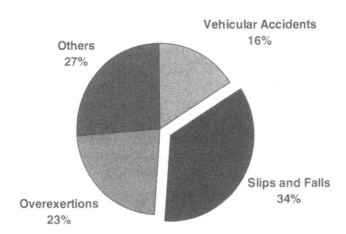

Fig. 5-11 Trucking industry 1997 workers' compensation claims of at least $1000.00 [7].

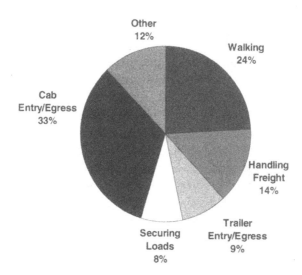

Fig. 5-12 Leading causes of trucking industry slips and falls [7].

and grab handles without exceeding the joint's normal range of motion. With physical accommodation satisfied, the stability of the individual ascending into or descending out of the vehicle can be evaluated by allowing three points of contact with the vehicle at all times and ensuring that the individual's center of mass remains within these contact points.

5.7 Vehicle Entry/Egress Analysis

During the initial phases of the entry/egress system design, the proper position of grab handles must be determined. Reach envelopes for each of the manikins can be generated to identify suitable areas on the vehicle where grab handles can be mounted. As seen in Fig. 5-13, the right-hand reach envelope of a small female manikin indicates that the lower portion of the B-pillar is reachable. Additionally, the left-hand reach curves (Fig. 5-14) indicate that the lower portion of the driver's door provides a suitable spot for a grab handle.

With these grab handle areas identified, prototype grab handles can be affixed to the door and evaluated and modified using the manikin's joint range of motion. In this case, the small female's left shoulder and hip angles exceed an acceptable range of motion [8]. Therefore, the interior door grab handle should be positioned lower on the door to reduce the shoulder abduction necessary for

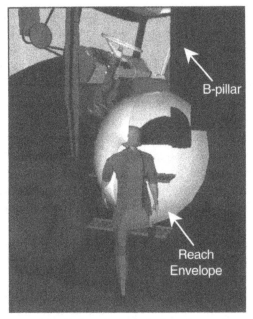

Fig. 5-13 Small female's right-hand reach envelope.

Fig. 5-14 Small female's left-hand reach envelope.

grasping the grab handle. Furthermore, the step height of this vehicle requires greater hip flexibility than is possessed by the majority of the U.S. population [8] and should be lowered.

To ensure stability while ascending into and descending out of heavy vehicle cabs, it is recommended that the individual entering or exiting the vehicle maintain three points of contact with the vehicle at all times during entry/ egress [3]. In other words, the person should have at least three limbs, two hands and one foot or one hand and two feet, contacting the vehicle throughout the ascent and descent. Digital human modeling software is ideal for demonstrating this simple task. Based on direct observation of actual people climbing into vehicle cabs, entry and egress strategies can be created representing the natural climbing behaviors of drivers. Then, these strategies can be used to guide the placement of the manikin hands and feet to reliably predict the individual grasp and step points. Although this process cannot account for everyone's unique natural behaviors or the impact of experience on climbing strategies and behaviors, the use of human modeling provides an indication that three points of contact are possible (Figs. 5-15, 5-16, and 5-17).

Fig. 5-15   Three points of contact with vehicle at first step.

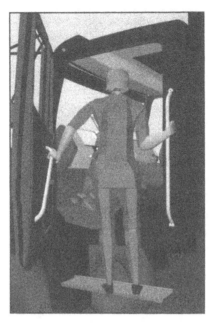

Fig. 5-16   Three points of contact with vehicle at second step.

Fig. 5-17   Three points of contact with vehicle into cab.

We can further examine this idea of three points of contact by using the estimated location of the manikin's center of mass. By tracking the change in position of the manikin's center of mass relative to the load bearing points of contact, we can get an indication of the relative stability of the individual as he/she ascends into or descend out of the vehicle. For instance, we can assume more equal distribution of body mass about the three points and greater stability when the manikin's center of mass is within these three points of contact. Conversely, we can assume a greater inequality of loading about the three points of contact and instability as the manikin's center of mass moves away from these points of contact. The results of these analyses indicated that center of mass moves farther away from the three points of contact as the manikin enters the vehicle (Fig. 5-18 and Fig. 5-19). Therefore, the greatest potential for instability occurs as the individual transitions from the step and into the vehicle cab (Fig. 5-20).

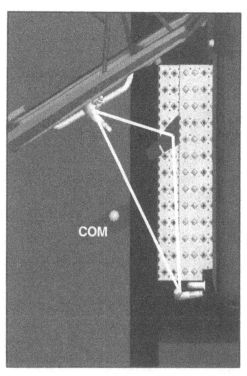

Fig. 5-18 Position of the manikin's center of mass (COM) at first step.

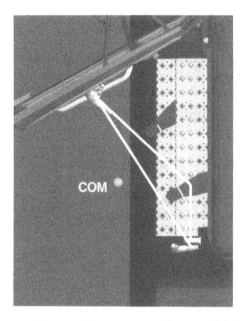

Fig. 5-19 Position of the manikin's center of mass (COM) at second step.

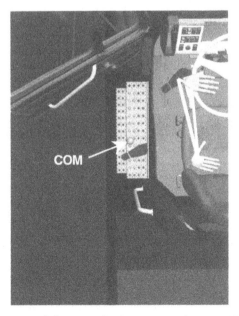

Fig. 5-20 Position of the manikin's center of mass (COM) into cab.

5.8 Advantages of the Technology

Although digital human modeling is still a technology early in its evolutionary cycle, the technology possesses numerous strengths that will perpetuate its existence in the product development environment. The primary strengths of digital human modeling include the efficiency and speed for conducting product analyses and system integration.

Due to the nature of the technology, the greatest advantage it provides a product development company is the decreased cycle time between concept development and functional validation. It is estimated, based on my experience, that the use of digital human modeling has reduced our product development analyses time by roughly forty percent. Of this forty percent, the majority of the time was consumed by translating and importing the computer-aided design (CAD) geometry data into the digital human modeling software. With improved software compatibility and improved means to generate virtual environments, I believe the time spent for product development analyses can be reduced by more than seventy percent.

In addition to the reduced product analyses times, digital human modeling software provides the capability to conduct quick "what if" analyses of design alternatives. The design of the vehicle can be quickly altered to determine optimum design configurations. With proper CAD engineering support, several unique virtual system configurations can be evaluated in a relatively short amount of time, as opposed to waiting for the vehicle system to be physically altered. Instead of providing the design engineers with evaluation results that indicate only whether the product is or is not ergonomically viable, the ergonomic engineer can use virtual vehicle environments to determine and report the optimum configuration specifics.

The software provides the opportunity to use standardized evaluation methodologies to systematically vary system characteristics or human capabilities to determine impact on design. Just as a scientist would use experimental methods to establish cause and effect relationships in research, the ergonomic engineer can use digital human modeling to systematically vary both the vehicle configuration and human characteristics to empirically investigate vehicle system design.

As stated by many sources in the current literature, digital human modeling diminishes the costs associated with traditional product development analyses [4, 5, 6] Much of the cost reduction in product analyses is due to the decreased need for physical prototypes. Therefore, the costs associated with the development of devices such as seating bucks and vehicle prototypes, can be reduced or eliminated. By using virtual human models, the need to recruit and pay real human participants is removed from the analyses. Finally, the use of digital human modeling reduces the need and costs for maintaining the facilities to conduct physical analyses.

In order to understand the human-machine interaction, the system level characteristics, such as component rotation and translation, are added to the vehicle model within the digital human modeling software to provide a means to virtually manipulate the vehicle-level systems. Therefore, the software provides one of the first opportunities to ensure the integration of a vehicle's subsystem components.

5.9 Disadvantages of the Technology

Despite the previously mentioned utility of the technology, digital human modeling has several limitations that can be overcome as the technology matures. The foremost disadvantages of digital human modeling that limit its utility in a product development environment include the novelty of the technology and lack of realistic psychomotor behaviors and motions.

The newness of the technology, or its novelty to industry, gives those individuals without a full understanding of the technology a false sense of its capabilities. Not understanding the underlying assumptions of these human models, some individuals receive the results of the digital human modeling analyses as the "absolute truth," when in reality the software analyses were extended into areas that violated the underlying empirical models. To overcome this, the person conducting the analyses must (1) be aware of this tendency; (2) ensure that the software is not used to examine issues that are outside the software's fundamental model assumptions; and (3) educate those who will receive the results of these analyses about the limitations of the current technology. There are two certainties of today's digital human modeling. First, the capability that this technology provides to simulate and illustrate the human-machine interaction early in the design cycle is powerful. Second, this simulated human-machine interaction will occur only in the manner the software user intends, whether these intentions are valid or invalid.

The other major disadvantage of the today's digital human technology is the lack of psychomotor behaviors incorporated into the manikin's motion analyses. Although efforts are being undertaken to provide digital manikins with realistic reach strategies, the current models move through unrealistic linear paths. Additionally, the current digital human models do not account for variations of human motion due to emergent behaviors that are the result of experience and/or adaptation to system characteristics.

5.10 Foreseeable Trends in the Technology

Based on present experiences with digital human modeling, the technology appears to be growing in three primary areas. The most accelerated growth for digital human modeling appears to be in the virtual reality and visualization realm. This trend is evident based upon the overwhelming interest of electronic gaming and animation industries to provide a realistic human form to improve the realism of virtual entertainment. Next, this technology is growing quickly in the industrial ergonomics area. Many of the current traditional tools (i.e., manual material handling limits, lower back analyses, etc.) available to industrial ergonomists are being incorporated into digital human modeling. Although slower than the aforementioned areas, there is also growth in the product development environments. This growth is evident by the incorporation of SAE recommended practices, traditionally paper-and-pencil exercises, into the digital human modeling software.

Another foreseeable trend in digital human modeling technology is the movement toward the software becoming more customer-specific. As the areas in which the technology is used become more diverse, the software companies' ability to provide enhanced software functionality in all these areas becomes more difficult. Therefore, clients of these software packages are required to either rely on software companies' consultation services to develop customer-specific software upgrades or to develop in-house capabilities to conduct proprietary research and implement custom software programming to overcome this dilution of the software functionality.

Still another anticipated expansion of the digital human modeling technology is in the area of human motion capture. By merging the motion capture system's ability to quantify human kinematics and the visualization capacity of human modeling, many human modeling applications (e.g., researchers, engineers, and animators) will be able to incorporate natural psychomotor behaviors into virtual environments with relative ease. In recent years, the applications of human motion capture to human digital modeling have been limited by software integration issues, environments within which motion captures can occur, and the ability to export real-time human motion data. However, recent advances in both software and hardware have enhanced the utility and compatibility of these technologies. As software companies begin to realize the importance of motion capture in human modeling, enhanced software is being developed with the functionality necessary to effortlessly import human capture data into human model environments. In addition, the motion capture industry has improved the hardware and software to expand both the type of environments (e.g., industrial, vehicular, etc.) in which data can be collected and the technologies (e.g., optical, magnetic, etc.) with the ability to capture and export real-time human motion. With these advances, it is anticipated that human motion capture will become a seamless extension of the human modeling system.

5.11 Conclusions

Throughout this chapter, the application of digital human modeling for demonstrating improved vehicle safety early in the heavy commercial vehicle design evolution is reviewed. The primary focus of this chapter is proper packaging of the driver's environment and vehicle cab entry and egress. In addition to examining the application of digital human modeling in a heavy commercial vehicle environment, the advantages and disadvantages of the technology are reviewed. Finally, this chapter presents emerging trends that may take place over the next few years within the field of digital human modeling.

As the product development community continually moves toward the use of virtual technologies to engineer and validate product designs, those individuals who evaluate the human-machine interface must find ways to bridge the gap between traditional assessment techniques and the emerging digital design and engineering tools. Digital human modeling has demonstrated over the past decade its increasing adeptness at filling this void.

References

1. Department of Transportation, Federal Highway Traffic Safety Administration, "Controls and Displays," 49 Code of Federal Regulations Part 571.101, September 1998.
2. Department of Transportation, Federal Highway Administration, "Step, Handhold, and Deck Requirements for Commercial Motor Vehicles," Federal Motor Carrier Safety Regulation Part 399 Subpart L, November 1998.
3. American Trucking Associations, The Maintenance Council, Truck and Truck Tractor Access Systems, Recommended Practice 404B, April 1989.
4. Ianni, J.; Clark, K.; Blaney, L.; Hale, R.; Ziolek, S.; and Bridgman, T., "Maintenance Hazard Simulation: A Study of Contributing Factors," in *Third Annual Symposium on Human Interaction with Complex Systems,* IEEE Computer Society, August 1996.
5. Rastetter, Ina, "Ergonomic Designs of Mercedes-Benz Trucks at DaimlerChrysler," SAE Paper No. 1999-01-3736, Society of Automotive Engineers, Warrendale, Pa., 1999.
6. Brown, Alan S., "Role Models: Virtual People Take On the Job of Testing Complex Designs," *Mechanical Engineering,* p. 44, July 1999.
7. McCullough, Patricia, "Insurance Handbook: Workers' Comp; Diligence, Safety, Back-to-Work Programs Keep Costs Down," *Heavy Duty Trucking,* June 1998.
8. Tilley, Alvin R., *The Measure of Man and Woman,* Whitney Library of Design, New York, N.Y., 1993.

Chapter 6

The Determination of the Human Factors/Occupant Packaging Requirements for Adjustable Pedal Systems

Deborah D. Thompson, PhD
DaimlerChrysler Corporation

6.1 Introduction

One of the most important aspects of designing vehicle systems is creating an interior compartment that is accommodating, or meets the needs of the customer population that the vehicle is targeting. If the design does not fully meet the requirements of the customer, it will probably fail to realize its full market potential [1]. Occupant and interior packaging involves determining the necessary amount of interior space for the environment and arranging interior and structural components to enhance the performance of the customer, in particular the driver [2]. The occupant and interior packaging process relies on the human factors and ergonomics approach to design for the customer. When employing human factors and ergonomics techniques, the objective is to design the environment to fit the capabilities and limitations of the customer by [3]:

- Identifying the target customer, and fully recognizing their needs and requirements for comfort, safety, and accommodation through understanding their physical, cognitive, and behavioral characteristics.

- Accommodating the entire range of the target customer population by designing the interior compartment to address their needs and requirements given their characteristics.

Packaging the occupant within the interior environment is a difficult task given the potential wide range of sizes, proportions, and skills of the occupant. However, applying appropriate occupant packaging and human factors concepts and methodologies can result in developing vehicle systems that will be successful in meeting the needs of the customer.

However, the challenges to achieving accommodation objectives are also constrained by the business objectives of the company. In order to survive in today's highly competitive global marketplace, many organizations, including DaimlerChrysler Corporation, have adopted business objectives such as:

- Reduced concept to market development ("speed to market")

- Lower development costs by reducing/eliminating costly physical prototypes and mockups

- Increased product quality

Many organizations are attempting to achieve these objectives by incorporating virtual prototyping technology into the development process to obtain a competitive advantage. At DaimlerChrysler, our focus in achieving these objectives has been through the incorporation of simulation technology. For example, the use of Digital Modeling Assembly (DMA) simulation technology assisted in taking eight months off of the product development process for our 1998 LH vehicle line (Dodge Intrepid, Chrysler Concorde, LHS and 300M). However, with the incorporation of these technologies, a dilemma exists: If the goal is to design cars and trucks that satisfy the needs and requirements of the customer, how do we simulate the customer within the product development simulation environment?

6.1.1 Use of Human Simulation Technology

A simulated human (hominoid) has a geometric representation, visual appearance, and articulation resulting in a good representation of an actual human [4].

This technology allows for the representation of the size, shape, and gender of a diverse customer population, thus providing the primary information to simulate the appropriate target customer group. The incorporation of human simulation technology within the product development simulation environment provides the ability to evaluate whether the packaging requirements of the customer are satisfied. Human simulation models allow for the evaluation of, and modifications to, the design without the expense and time that would be incurred through the development of a physical mockup/prototype to physically simulate the interior vehicle environment. Therefore, human simulation technology provides a means to determine if the design satisfies the customer, while adhering to the business objectives of the company.

6.2 Problem Analysis: Accommodating the Target Customer Population

During the early stages of the product development process, the target customer is identified. Based on the niche/class of the vehicle (compact, midsize, large car, truck, etc.), the target customer is the individual who would tend to purchase the vehicle. The consumer research activity that is conducted to determine the target customer results in a profile that indicates their potential needs, preferences, and expectations of the vehicle. This profile details the characteristics of the customer, indicating their attributes such as age, gender, educational level, income range, and lifestyle, which can be used to determine the customer's capabilities and limitations that can impact the design of the vehicle [3].

Once the customer population has been identified, the next objective is to develop designs that appropriately accommodate them and addresses their needs. However, current industry interior design paradigms have resulted in vehicles that are basically designed to accommodate males, in particular large males, and therefore disenfranchise many including the young, the elderly, the disabled, and women [5, 6]. For example, one of the most important factors that influences the choice of a seating position in the vehicle is the ability to comfortably reach the pedals. Due to industry paradigms, smaller female drivers, in order to reach the pedals of the vehicle, must sit close to the steering wheel [7]. The male-dominated corporate culture of the automotive industry, coupled with the tendency to design per intuition has resulted in vehicle designs that do not reflect the differences between males and females, leading to a lack of accommodation for females [8]. This fact has been confirmed by the industry itself [9]. Until the industry improves vehicle designs to accommodate a more diverse customer population of varying sizes, shapes, etc., short-term solutions such as adjustable pedals have been investigated.

Adjustable pedals provide the ability to adjust the pedals by providing movement of the brake, accelerator, and clutch assembly (Fig. 6-1). OEMs, and suppliers of adjustable pedal systems have touted their benefits including [10, 11]:

- Improving comfort for smaller stature drivers
- Optimizing driver seating position.
- Reducing seat track travel

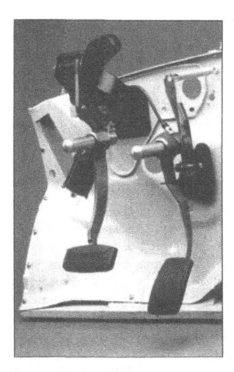

Fig. 6-1 Example of power adjustable pedal system. (System by Teleflex, Inc. for the '99 Ford Excursion and Lincoln Navigator[10].)

In 1992, the former Chrysler Corporation introduced the Dodge Viper, the first production vehicle to provide adjustable pedals. (The 1999 Lincoln Navigator and Ford Excursion are the first production vehicles to provide power adjustable pedals, as the initial Viper system was manually controlled). In investigating the adjustable pedal system option for our vehicles, DaimlerChrysler recognized that there were no internal design requirements for packaging the pedal system. Although the system was packaged in the Viper, the sports car "niche" of the vehicle, coupled with its low production volume does not result in the type of large, diverse customer population that would occur with vehicles with a larger market appeal. Therefore, a strategy was developed whereby our design requirements could be generated for vehicles with large production volumes, and provided to our Extended Enterprise™ of suppliers.

6.3 Design Analysis: Packaging Requirements for Adjustable Pedal Systems

6.3.1 Set-up for the Analysis

A previous investigation into the types of human simulation technology available resulted in the decision to utilize SAFEWORK, developed by Genicom Consultants, Inc., of Montreal, Canada (Fig. 6-2). However, it had to be determined if human simulation technology could be employed to develop the design requirements for the adjustable pedal system. In packaging the driver, a seated body posture is required for normal driving operations. However, the body posture that is selected by the driver is a function of his/her physical characteristics (anthropometry, somatotype, static and functional capabilities) and their perceptions (how they feel, are they comfortable, etc.), so the design requirements that are developed must reflect these attributes. It was found that human simulation technology has evolved so that it can be used to effectively simulate physical characteristics; however, it has not reached the level of sophistication to predict how the human perceives [12, 13]. In order to gauge human perceptions, an evaluation study would have to be conducted, requiring the development of a physical mockup/prototype to simulate the interior vehicle environment, thus incurring expenses and time, which is contrary to our business objectives. This barrier was overcome by employing human factors/ergonomics research data concerning the actual postures that individuals adopt in driving situations, and

Fig. 6-2 Example of SAFEWORK within an interior vehicle environment. (Courtesy of Genicom Consultants, Inc.)

relaying the body posture angles which incorporate the physical characteristics and perception attributes. A number of research sources were reviewed [14, 15, 16, 17, 18], resulting in the adoption of the following accommodation posture envelope for our design analysis:

Body Posture	Angle (degrees)	Design Criteria
Back	95–100	alertness
Elbow	80–165	comfort
Upper Arm	0–35	comfort
Knee	110–120	strength
Ankle	90–100	comfort

A 5th percentile, in stature, female hominoid was used to perform the initial analysis, given the current lack of accommodation for females, and the touted benefits of adjustable pedal systems in improving comfort for smaller stature drivers such as females, who, on average, are shorter than males [19, 20]. The anthropometric database on which the hominoid was based was the 1988 Natick anthropometric study [21], which was chosen in order to utilize the most recent anthropometric information available. Since the Natick database is based on the measurement of military personnel, whose selection criteria result in the exclusion of many individuals, some may question if the military is representative of the customer population for whom we design vehicles. A study by McConville, Robinette, and Churchill [22] found that military data could be used to determine the data of the general population since approximately 94–99% of the civilian data could be matched with military individuals, with a good fit for stature and other linear dimensions, but not necessarily for depth and breadth dimensions. Based on these results, we were comfortable with using the military anthropometric database to simulate the basic linear dimensions of our customer population.

The SAFEWORK software was used as a stand-alone package, and was not integrated in the computer-aided design (CAD) system, called CATIA (Computer-Aided Three-Dimensional Interactive Application), that DaimlerChrysler utilizes for its product development activities. DaimlerChrysler is concerned about the security of its product design concepts and has made it extremely difficult to export a design geometry from CATIA to other software packages. This presented a challenge to our design analysis that made it necessary to duplicate the basic geometry of the vehicle package in SAFEWORK. In order to expedite the study, only the basic primary controls (pedals and steering wheel), along with the seat and seat track travel range, were generated in SAFEWORK.

The result of positioning the 5th percentile female hominoid, with her dimensions based on the 1988 Natick anthropometric database, per the appropriate accommodation body posture angles, is shown in Fig. 6-3. With regard to accommodation angles, the left side of the hominoid represented the higher end of the accommodation range, while the right side represented the lower end. With the hominoid positioned so that she could reach the pedals, the figure provides a graphic illustration of the lack of accommodation that was detailed earlier.

Fig. 6-3 Set-up of design analysis of SAFEWORK 5th percentile female hominoid within basic design geometry.

6.3.2 Result of the Analysis

In order to provide the appropriate accommodation, the hominoid was moved rearward, along the seat track travel line, until the steering wheel was in between the accommodation ranges for the arms (Fig. 6-4). This made it necessary for the pedals to be moved rearward, a minimum of 76.2 mm, to accommodate the movement of the hominoid. Also, since the seat track in our vehicles is typically at a 7.5 degree angle, as the hominoid was moved rearward, she also moved down in the vertical direction. Therefore, some downward movement of the pedals was also necessary. The analysis suggested that a downward movement a minimum of 50.8 mm would be necessary in order for the ball of foot of the hominoid to be in contact with the pedals. So from

Fig. 6-4 Results of analysis to define the packaging requirements of the adjustable pedal system.

the nominal design position of our pedal package, the adjustable pedal system required a minimum travel range of 76.2 mm rearward and 50.8 mm downward.

6.4 Conclusion

The use of human simulation technology was successful in allowing us to determine the packaging requirements for the adjustable pedal system, while adhering to our business objectives. The time required to perform an evaluation study (approximately three to four months) was dramatically reduced (to approximately one day) through use of the technology. Had it not been necessary to duplicate the basic package geometry into SAFEWORK, the time to perform the analysis could have easily been reduced to a few hours. However, without the use of human factors/ergonomics data, this analysis could not have been performed at all, since the technology currently is not capable of predicting human perceptions. Also, since only the basic package geometry was used, the analysis was not able to address the impact of adjustable pedals on other aspects of the design of the vehicle. For example, adjustable pedals provided the ability to move the smaller female more rearward in the vehicle and provide better accommodation. But with the 7.5 degree seat track angle, which causes the driver to move down in the vehicle as the seat is moved rearward, the question arises as to whether this would have a negative impact on the ability of the smaller female to see over the hood (a reduction in her down visibility angle) in order to have good "command of the road" feel.

The developers of human simulation technology are increasingly aware of the need to develop software that fits the needs of their end users. In particular, since this analysis, Genicom has entered into a strategic partnership with Dassault Systems, the developers of CATIA. Together they are developing an application architecture so that SAFEWORK will operate within CATIA [23]. So, although there are currently shortcomings associated with the technology, as the benefits of using the technology are realized, more demand, and capital, will result in its continued evolution.

References

1. Pugh, S., *Total Design*, Addison Wesley Publishers Ltd., England, 1990.
2. Roe, R.W., "Occupant Packaging," in Peacock, B. and Karwowski, W. (Eds.), *Automotive Ergonomics*, pp. 11–42, Taylor & Francis, Washington, D.C., 1993.
3. Cushman, W.H. and Rosenberg, D.J., *Human Factors in Product Design*, Elsevier, Amsterdam, 1991.
4. Granieri, J.P. and Badler, N.I., "Simulating Humans in VR," Technical Report - Center for Human Modeling and Simulation, University of Pennsylvania, Philadelphia, 1994.
5. Porter, J.M. and Porter, C.S., "Turning Automotive Design Inside-Out," *International Journal of Vehicle Design*, Vol. 19, No. 4, pp. 385–401, 1998.
6. Freund, P. and Martin, G., *The Ecology of the Automobile*, Black Rose Books, Montreal, 1993.
7. DeLeonardis, D.M.; Ferguson, S.A.; and Pantula, J.F., "Driver Seating Position Survey," *Automotive Engineering International*, pp. 69–72, May 1998.
8. Thompson, D.D., "An Ergonomic Process to Assess the Vehicle Design to Satisfy Customer Needs," *International Journal of Vehicle Design*, Vol. 16, Nos. 2/3, pp. 150–157, 1995.
9. Yung, K., "Ford Designers Pay More Attention to Women's Needs," *The Detroit News*, July 12, 1998: C (http://detnews.com/1998/autos/9807/12/07120020.htm).
10. http://www.tfxauto.com, updated 1999, Teleflex Automotive Group, Teleflex, Incorporated.
11. Automotive Industries Staff, "Components and Systems 1999: Highlighting the Hottest New Hardware for the '99 Model Year," *Automotive Industries*, p. 41, December 1998.
12. McDaniel, J.W., "Human Modeling: Yesterday, Today, and Tomorrow," SAE International Conference, Digital Human Modeling for Design and Engineering, Dayton, Ohio, 1998.
13. Badler, N.I.; Phillips, C.; and Webber, B., *Simulating Humans: Computer Graphics Animation and Control*, Oxford University Press, New York, 1993.

14. Porter, J.M. and Gyi, D.E., "Exploring the Optimum Posture for Driver Comfort," *International Journal of Vehicle Design*, Vol. 19, No. 3, pp. 255–266, 1998.
15. Henry Dreyfuss Associates, *The Measure of Man and Woman: Human Factors in Design*, Whitney Library of Design, New York, 1993.
16. Grandjean, E., "Sitting Posture of Car Drivers from the Point of View of Ergonomics," in Grandjean, E. (Ed.), *Human Factors in Transport Research (Part 1)*, pp. 20–213, Taylor & Francis, London, 1980.
17. Preuschen, G. and Dupuis, H., "Posture and Seat Design for the Car Driver," Proceedings of the Symposium on Sitting Posture, pp. 120–131, 1969.
18. Rebiffe, R., "Ergonomic Investigation of the Arrangement of the Driver's Seat in Private Cars," IME Symposium, 1966.
19. Pheasant, S., *Bodyspace: Anthropometry, Ergonomics and the Design of Work*, Taylor & Francis, London, 1996.
20. Kroemer, K.H.E.; Kroemer, H.J.; and Kroemer-Elbert, K.E., *Engineering Physiology Bases of Human Factors/Ergonomics*, Van Hostrand Reinhold, New York, 1990.
21. Gordon, C.C.; Churchill, T.; Clauser, C.E.; Bradtmitter, B.; McConville, J.T.; Tebbetts, I.; and Walker, R.A., "1988 Anthropometric Survey of US Army Personnel: Methods and Summary Statistics," Technical Report, NATICK/TR-89-044, United States Army Natick Research Development and Engineering Center, Natick, Mass., 1989.
22. McConville, J.T.; Robinette, K.M.; and Churchill, T., "An Anthropometric DataBase for Commercial Design Applications," Final Report, NSF DAR-80 09 861, Anthropology Research Project, Inc., 1981.
23. http://www.safework.com, updated 1999, SAFEWORK, Genicom Consultants, Inc.

Chapter 7

Ergonomics Analysis of Sheet-Metal Handling

Brian Peacock, Heather Reed, and Robert Fox
Manufacturing Ergonomics Laboratory
General Motors Corporation

7.1 Background

Sheet-metal handling is a major challenge for ergonomists. It represents an application to which various analytical and simulation methods can be applied. These methods range from simple checklists through time studies and biomechanical analyses to anthropomorphic modeling. Like other manufacturing and assembly operations, the design of sheet-metal handling equipment and operations have many constraints. For example large, capital-intensive equipment, such as transfer presses, must be used efficiently. Also, the production demands by the assembly plant are such that a timely supply of components is essential. Parts must not be damaged in transit; therefore containers may be designed more for protection of the part than ease of getting the part in or out. The production demands may be such that runs of a particular part may only last a few hours before the dies are changed for a different panel. These varying products of the stamping process require the versatility of a human operator for end-of-line off loading. In contrast, the assembly plant body shop will typically be set up so that a particular station handles the same part for many years, thus making automation more feasible and economical. The ergonomist, therefore, is faced with a complex set of constraints, as he or she tracks a day in the life of a sheet-metal stamping.

113

7.2 The Process

The typical manual sheet-metal handling demands are identified by an overview of the process as the metal is moved from the steel mill to the body shop for welding. Sometimes welding is carried out in the stamping plant and sometimes in the assembly plant. A roll of sheet metal arrives from the steel mill and is cut into blanks, of appropriate size for the required panel, which are placed at the beginning of a transfer press line. Sometimes the blanks are cut before delivery to the stamping plant. Because of the size and weight of these rolls and blanks and the need to minimize scrap, these activities are usually carried out by hard automation (automatic materials handling equipment). Typical components will include a floor pan, hood, roof panel, side-frame, door, and fender.

The blanks are fed automatically into the press line where they are moved from die to die to form the desired shape of the part. The first manual station is usually at the end of the press line where the panels are visually inspected and off-loaded into part specific containers. The orientation of the part in the container will be dependent on a variety of things including: size, type of surface, the need for protection during transportation, and other process requirements. The larger parts may be oriented on their edges in dunnage that holds them in place during transit. Alternatively, stampings may be placed horizontally with packing in between panels to maintain their shape. Smaller parts may be stacked together and placed in a horizontal orientation.

The production demands and the capacity of the capital intensive presses may be such that the panels pass through the transfer press line at up to forty per minute, depending on the size of the part. Once a metal container is full, say with twenty large panels, they are removed by fork truck to the rail car, loading dock, or to a welding area within the stamping plant. It should be noted that the rapid throughput rate requires that containers must be arranged at the end of the press line so that one may be filled while another is removed. There are a variety of arrangements of containers and off-loading stations at the end of the press line, but typical arrangements include semicircles of metal baskets or longer conveyor belts with parallel lines of metal containers. Because of the mixture of batches of panel types coming off the press line, it is common to employ manual methods at this station. It should be noted at this juncture that the breaks in flow through modern press lines for die changes are now only a few minutes.

The panels are then transported by rail or road to the assembly plant. Here the larger panels are transferred in their metal containers directly to the body shop for welding. Smaller panels may undergo a repacking or kitting process, either in the stamping plant or in the assembly plant. This activity entails placing groups of different panels into a large container with a sufficient supply for say one hour's production in a body shop work cell. A typical vehicle assembly throughput rate is sixty per hour, which contrasts sharply with the throughput of up to forty per minute in the stamping plant.

In the body shop, the materials delivery operator will place a stack of small parts, sufficient for an hour's production, conveniently for the production operator. Commonly the welding machines are arranged around a horseshoe shaped cell with alternating parts containers and welding fixtures. The cell operator will walk around the cell loading and off-loading the increasingly built-up body component. Where the work cell is not appropriately designed, it is common to see the production operator relocate or "stage" some parts for convenience of loading into the welding fixture. The baskets containing larger panels are sometimes placed for the convenience of the production operator, but their location is sometimes constrained for the ease of access by the fork truck operator. Again staging of such parts as rails—long thin components—is not uncommon. Of course in robotic welding cells the materials operator will be required to place the containers much more precisely.

The containerization design group is often challenged with the task of meeting conflicting demands of the metal fabrication plant and the body assembly shop in the vehicle assembly plant. It is not uncommon for the metal fabrication plant to have a requirement that the parts be loaded into the container manually, while the body shop may be designed for robotic loading of the welding fixtures. In the stamping plant the human operator's versatility is needed for the mixture of parts and containers that are processed within a day, whereas in the assembly plant the same part may be used for the lifetime of a product, which may be many years. The decision between manual and automatic component handling is based on many factors such as human versatility, floor space, the size and weight of the part, and the length of a run of a particular part. On occasion there may be manual backup arrangements for robotic stations. The problem for the container designers is that the requirements for the robot are usually different from those of the human operator. Finally, because of the confined access to the various areas it may be necessary for human operators to move large containers on wheels.

7.3 Outcomes

The ergonomics approach to sheet-metal handling, like most other industrial activities, involves first an assessment of the desirable and undesirable outcomes of the process. The desirable outcomes are product quality and productivity. The product quality outcome is achieved by designing all stages of the forming and handling process so that specifications are met and damage does not occur. The interjection of various inspection and recycle or repair functions helps to assure this quality requirement. The productivity requirement involves a balance of the throughput of all the stages so that the demand of the later stages may be met. This throughput must also make full use of available resources, including material handlers. Thus the design of workplaces, containers, and jobs must be compatible with the requirements of these human material handlers. Because the presses are often capable of producing up to 40 panels per minute, whereas the demands of the body shop will typically be on the order of 60 panels per hour, it is imperative that precise workload analysis be conducted to meet these production demands. This workload analysis will be used to estimate the number of operators needed at each handling stage, with due reference to the physical demands of the tasks.

The undesirable outcomes of the handling process are damage to the part and acute and cumulative injury to the operator. Parts may suffer dings, scratches, and deformations by inappropriate handling. The typical acute injuries to the operators include lacerations and abrasions. The cumulative illnesses include a wide variety of musculo-skeletal disorders associated with the backs, shoulders, elbows, wrists, and hands of the operators. Because the activities are often highly repetitive, it is essential that the physical arrangements and job requirements are addressed so as to eliminate or substantially reduce the incidence of such disorders. It should be noted at this juncture that personal and other factors may also cause musculo-skeletal disorders to the extent that attention to workplace and task design may not always be sufficient to eliminate them altogether.

7.4 Decisions

Ergonomics analysis of the sheet-metal handling jobs will assess their physical and temporal characteristics with the purpose of optimizing the various outcomes. The ergonomist will make suggestions regarding the design of workplaces, containers, handling assist devices, automation, and work assignments. The engineering and supervisory communities may implement these suggestions. However, before a particular intervention is implemented a management decision will be made with regard to the costs and benefits of the alternative designs. The overall production rates in the stamping plant and container designs will be dictated by the demands of the assembly plant body shop processes, with due regard to maintaining inventory at minimal levels while assuring sufficient supply to satisfy the assembly process requirements. It should also be noted that the high capital costs of stamping plant equipment and body shop robots is such that these processes should run continuously, at full capacity, with as little down time as possible.

7.5 Design

Variables that are amenable to manipulation include the spatial characteristics of the stamping and assembly workplaces, including heights, orientations, reaches, and access pathways. Because of the nature of the parts, the layout must include provisions to prevent collisions between parts, equipment, and people. Required forces and "targets" may also be optimized. The workplace may also include devices to orient the panels in a way that is conducive to grasping and transfer. A variety of handling aids, ranging from simple hoists and articulating arms to semi-intelligent devices that feature vertical assistance and pathway guidance may be used. End of line containers may be oriented and presented in such a way as to minimize the movement distances and reorientations of the panels. Slots may be arranged to provide friendly targets and so obviate the need for multiple adjustments in the placement, positioning, and removal processes. Welding fixture locations, orientations, and guidance features can be manipulated to facilitate the manual transfer process. Full automation is also an option under certain circumstances. Where manual operations are chosen, decisions must be made regarding the staffing levels and job assignment patterns of the operators. Job rotation strategies are often employed to combat the highly repetitive and monotonous nature of the manual tasks and to optimize the human contributions.

7.6 Choices

The choice between alternative designs is not always straightforward and may require sophisticated ergonomics analysis. First, the decision whether or not to automate the whole process is not always feasible because of the variability in demand and the inherent reliability and versatility of the human operators. Labor agreements may also contribute to the decision. Some automatic processes cannot meet the throughput requirements. More complex automation to match the versatility of human operators may not be available or cost effective, although there has been a trend toward greater levels of automation in the past two decades. Handling devices are used for heavy parts, but not necessarily for parts, say, below 40 pounds. The typical reason offered for not using these devices is that the operator may be able to work more quickly without them, thus creating more discretionary time. Another complicating factor is the self selection by operators that typically choose to work in stamping plants and body shops. Handling devices, such as rails and bridges with hoists also may require more floor-space than manual operations and space is often a premium resource. Handling devices generally only deal with the gravitational load and may add to the inertial load of the handling task. Contemporary intelligent assists, which address gravity, inertia, and guidance may have some application in sheet-metal handling, but to date they have not proved to be as fast or agile as the manual process.

7.7 Analysis

The initial analysis technique that is applied to the various manual and automated sheet-metal handling processes is method and time measurement. Predetermined times are used to establish initial staffing levels. More detailed ergonomics analysis techniques are often superimposed on these basic work measurement methods. For example, ergonomics checklists may ask qualitative or quantitative questions regarding the physical parameters of the task— such as weights, sizes, shapes, clearances, distances, locations, orientations, pathways, moments, movements, targets, and interfaces. Anthropometric information will be brought to bear on the workplace spatial arrangements such as heights, reaches, and access routes. Biomechanical analysis will be used to further investigate the forces, moments, motion patterns, and inertial control requirements of the various tasks. Extensions of Fitts' Law [1] may be applied to review all the important target characteristics for the placement of panels in baskets or welding fixtures. Energy analysis [2] may be appropriate where the loads, distances, and lifting and carrying demands are high. Psychophysical data [3] may be used to establish lifting frequencies, and the NIOSH Lift Equation [4] (either the 1981 or 1991 version) has become well established as an important analysis tool in such jobs. Finally, upper limb activity analysis [5, 6] methods may be used to add detail to the finer movements involved in sheet-metal handling. Traditionally these methods are used in a hierarchical fashion. If a checklist approach indicates some uncertainty in the design decision, then specific secondary analysis tools may be brought to bear on the problem. (Note: The 1981 version of the NIOSH lift equation is more sensitive to spatial and load factors, whereas the 1991 version is more sensitive to frequency factors.)

The more sophisticated anthropometric and biomechanical modeling approaches also have potential applications, both in the analysis of existing workplaces and methods and the design of new ones. The University of Michigan's Three-Dimensional Static Strength Prediction Program (3DSSPP) may be used to model various anthropometric, postural (joint angle), joint torque, and spinal segment force scenarios [7]. This modeling facility includes various ways of creating a desired posture by inputting joint angles, dragging joints, or by an inverse kinematic posture prediction approach, given the input of hand position. The program produces a plethora of analytic outputs, including the percentage of the population capable of handling a torque around a particular joint. In addition, it measures the tensions in various muscle groups and the compression, shear, and torsional forces in the spine. The University of Waterloo's "WATBACK" model performs similar analyses. These models are, however, limited to static analysis and thus do not deal with the movement and inertial matters that are important in sheet-metal handling. Given this shortcoming, these models are very useful in the simulation of alternative static scenarios.

7.8 Simulation

There are many "comprehensive" computer packages that may be used to extend the capability of the ergonomist in the analysis of the complex conditions found in sheet-metal handling. The proprietary names include SAMMIE, Jack, DENEB, RAMSIS, SAFEWORK, and ROBCADMAN. These models are basically anthropometric, that is, they allow manipulation of body shape and size, and joint angle in the context of a given workplace. In addition they have some kinematic features that allow the simulation of movements such as walking and materials handling. However, to date these movements are synthetic in form and do not mimic naturalistic human movement. Most of these models have additional features that are connected but not integral to the model. These features include timed movements linked to predetermined times, the NIOSH Lift Equation, the Liberty Mutual (Snook) Tables, the Garg Energy Model and RULA. The NIOSH Lift Equation allows the simulation of alternative static analyses at the beginning and end of a lift and the weighted averaging of multiple lifts. The Liberty Mutual Tables allow the analyst to estimate the proportion of the population capable of carrying out various manual materials handling activities, such as lifting, lowering, carrying, pulling, and pushing. RULA (Rapid Upper Limb Analysis) purports to assess the overall stress associated with a materials handling task, and the Garg Energy Model predicts the overall and elemental energy requirements of a sequence of activities associated with manual materials handling. A particular feature of the primary anthropometric model and the additional features is the ability to apply "flags" or decision thresholds. These thresholds may take the form of percent of the population capable or cutoff values based on some consensus of stress level. Research is underway at the University of Michigan to increase the level of integration of these various analytical and simulation features within a basic anthropomorphic model, starting with the incorporation of statistically based naturalistic human motion.

The high frequency of end of press line materials handling necessitates highly repetitive motions on the part of employees, while they are actually working. Regular breaks, equipment down time, rotations, and die changes create relief opportunities. An assessment or evaluation method that comprehensively addresses the consequences of repetitive motion is lacking from the arsenal of ergonomics tools. Such a tool would need to be epidemiologically based and go beyond the simplistic simulation uses of RULA in highlighting some out of range joint angles and stressful postures. In the discussion of the various tools it is germane to mention that the analysis results concerning human capabilities do not necessarily have a direct relationship to either injury or performance outcomes. The myriad assumptions and constructs concerning anthropometry, strength, pysychophysical capacities, etc. complicate the simulation user's understanding of risk. Human modeling and simulation provide a very powerful tool in the hands of the analyst, although the choices and inputs for the ergonomics analysis may be complex and not readily apparent to the untrained analyst. Thus the introduction of these powerful tools must be accompanied by appropriate organizational design and user education.

Simulation may assist metal fabrication and container engineers to make sound proactive choices in assigning products to appropriate press lines and in making appropriate workplace and equipment design decisions. With some exceptions, many pressed metal products are capable of running on a number of press lines and these decisions are made with regard to various factors, including production schedules. Sometimes the end of line arrangements may not be ideal for certain products. Simulation provides a tool that can rapidly help with end of line design and arrangement of containers and ancillary equipment.

7.9 Policy

These analytical and simulation tools provide very useful ways of assessing the complex human demands associated with sheet-metal handling tasks. In all cases, however, certain policy assumptions must be made to aid in the interpretation of and decisions based on the various tools. For example it is common to assume 5th and 95th percentile statures when assessing reach and fit situations. In biomechanical analysis, assumptions must be made regarding the size of the individual being modeled because of the effect of this feature on subsequent kinetic assessment. In this context it should be noted that taller individuals are generally stronger than shorter ones, but taller individuals have longer limbs and therefore possibly develop longer moment arms than shorter people. Consequently, it is critical that analyses are prefaced by certain fundamental assumptions. The 5th and 95th percentile policy assumption in anthropometry is somewhat simplistic because of the complex variability of segment lengths. This problem is compounded by decisions related to percentage of the population capable of certain joint torques or materials handling capabilities. These latter values are obtained from biomechanical and psychophysical studies which may or may not be representative of the anthropometric underpinnings. Thus the policy decision to "protect," say, 90% of a population begs the questions of which population, the technical derivation of the values, and the epidemiological importance of the protection level.

In the case of energy analysis, many assumptions must be made in the simulation that account for gender, age, weight, and conditioning over and above the basic anthropometric inputs to the analysis. When one considers the complex interactions between these human and task factors, it is clear that the policy decisions are at the same time important and difficult. The application of these individual and integrated models to the complex sequence of tasks associated with sheet-metal handling is therefore dependent on the input assumptions, policy, and the validity of the analytical equations. These difficulties are further complicated when it is realized that the actual modeler may have only minimal training in ergonomics, and therefore may not understand the complexities identified in the foregoing paragraphs.

7.10 Case Study

One particular example of the application of ergonomics analysis and modeling to the sheet-metal handling process is the design of a rack for a truck roof outer panel. The assembly plant made the decision to unload the part using a robot, and the metal fabrication plant prefers to load the roof to the rack manually. Because the groups involved did not necessarily work in parallel, the initial design of the rack was dictated by the requirements of the robot, and was designed by a separate containerization group. In this case the roof must be presented to the robot in "car position," meaning they are placed in the rack flat, one on top of the other, with separators. Table 7-1 contains the specifics of the part and rack-loading task at the manual station in the metal fabrication plant.

Table 7-1 Part (roof outer panel) and rack data

Part Weight	Approx. Part Dimensions	Proposed Frequency	Parts Per Rack	Rack Dimensions	Vertical Heights of Parts in Rack
21.56 lb.	W: 44.5 in. L: 59.1 in.	16 parts per min	20	W: 57 in. L: 96.5 in. H: 54 in.	6 to 45 in.

These racks are designed to be stacked on top of each other for transport and storage. Although the last roof loaded to the rack only reached a vertical height of 45 inches, the corner rails that support the racks placed on top were designed to be 54 inches high. This is to allow sufficient clearance between racks so that the forks of the fork truck can pass without doing any damage. As the roofs are placed in the rack, the operator flips down a placeholder on the supports at the four corners to keep the parts in place, and prevent them from laying directly on top of each other. This would create a nesting situation which would interfere with robotic removal in the assembly plant. The design also allows the robot at the body shop, through its force pulling up on the roof, to flip the placeholders back into the open position as the part is removed from the rack.

Once these requirements were developed based on the needs of the robot and the part size, the container design department asked for ergonomics advice regarding the ergonomics issues associated with loading the part at a frequency of 16 per minute. The interaction of the part size and weight indicated that two-person teams would be preferable. Because of the rack design, these teams would need to lift each part over the 54-inch rack corner supports. Traditional NIOSH lifting and energy expenditure calculations were carried out, with consideration of the limitations of the results because of the two-person team approach. The results were borderline (i.e., the analytic outcomes approached a level that could give rise to an increased incidence of musculoskeletal disorders), even when the work was divided between two teams of two people each (8 parts per minute per team). Also, the metal fabrication engineers knew from experience that lifting the part over the rack corner supports would be awkward, and could pose problems for smaller workers, regardless of the other risks involved.

At this point, a very simple model was set up using Engineering Animation, Inc.'s (EAI) Jack human modeling software. Within an hour, a novice modeler was able to generate an adequate representation of the roof rack and container using the software's built-in drawing tools. Next a 95th percentile male and 5th percentile female were added to simulate how this "best and worst case" team would maneuver a roof over the rack supports, and down into place in the rack. The simulation clearly showed the difficulties a smaller worker or workers that were mismatched in size would have navigating the rack supports, without reference to any additional biomechanical limitations. The model also exhibited the awkwardness that this design would pose to any pair of workers assigned to this task (see Figs 7-1 and 7-2).

*Fig. 7-1 Jack model of size mismatch in rack loading.
Note overhead work.*

*Fig. 7-2 Jack model of size mismatch in rack loading. Note back
bending for taller operator.*

The Jack software's analysis toolkit was also activated for further evaluation. The analysis modules included the NIOSH Lift Equation, the University of Michigan's Three-Dimensional Static Strength Prediction Program (3DSSPP) to assess joint torque and strength capabilities, an energy expenditure evaluation based on the Garg Model, and the Rapid Upper Limb Assessment (RULA) to evaluate the upper body stresses. Although the traditional NIOSH, RULA, and energy calculations were not very convincing, this further analysis and visualization (3DSSPP, EAI Jack) was helpful in supporting the argument that this rack design was far from ideal.

The metal fabrication engineers, with this ergonomics support, were successful in making their point that allowing the robot unloading process in the body shop to dictate the manual loading process was a poor decision, but one that they were forced to overcome. They would typically prefer to load the roof into the rack on its edge, eliminating the need to lift the panel over the corner support posts. This simulation stimulated discussion of a complete rack redesign, or possibly moving this part to another press line where the racks would be loaded semi-automatically. It should be noted that late design changes, which are not uncommon, have far reaching implications and costs. Consequently an ergonomics investigation, using simulation early in the design process drew attention to a potentially unsatisfactory and long lasting manual situation.

References

1. Fitts, P.M., "The Information Capacity of the Human Motor System in Controlling the Amplitude of Movement," *Journal of Experimental Psychology*, Vol. 47, pp. 381–391, 1954.
2. Garg, A.; D.B. Chaffin; and G.D. Herrin, "Prediction of Metabolic Rates for Manual Materials Handling Jobs," *AIHA Journal*, Vol. 39, p. 661, 1978.
3. Snook S.H. and V.M. Ciriello, "The Design of Manual Handling Tasks: Revised Tables of Maximum Acceptable Weights and Forces," *Ergonomics*, Vol. 34, No. 9, 1991.
4. Waters, T.R.; Putz-Anderson V.; Garg, A.; and L.J. Fine, "Revised NIOSH Equation for the Design and Evaluation of Manual Lifting Tasks," *Ergonomics*, Vol. 36, No. 7, 1993.
5. McAtamney, L. and E.N. Corlett, "RULA: A Survey Method for the Investigation of Work-Related Upper Limb Disorders," *Applied Ergonomics*, Vol. 24, No. 2, pp. 91, 1993.

6. Moore, J. and Garg, A., "The Strain Index: A Proposed Method to Analyze Jobs for Risk of Distal Upper Extremity Disorders," *American Industrial Hygiene Association Journal,* Vol. 56, pp. 443–458, 1995.
7. Chaffin, D.B., "Development of Computerized Human Static Strength Simulation Model for Job Design," *Human Factors and Ergonomics in Manufacturing,* Vol. 7, No. 4., pp. 305–322, 1997.

Chapter 8

Digital Human Modeling for Improved Product and Process Feasibility Studies

D. Glenn Jimmerson
Ford Motor Company

8.1 Introduction

The process of producing a transportation vehicle is composed of many subprocesses that must be accomplished in concert. These subprocesses include: identification of a customer product requirement; development and evaluation of a product design; development of a manufacturing process; design of specific equipment; design of the plants and facilities; and design of the logistics to move materials and products. Historically, each of these subprocesses required some type of prototype or model to prove out the concepts proposed. And, historically, each of these subprocesses was conducted with a minimum of interface and information exchange among the disciplines involved.

The nineties was a decade marked by extensive use of computers by the individuals involved in vehicle production, and this technological advancement enabled greater analysis and sharing of information. This technological explosion presented the opportunity for designers to create virtual models and share them, instantaneously, with others around the world.

Product development engineers, working with market research data, could take a current vehicle design and modify it to satisfy the requirements of a certain demographic group. If, for example, market research indicated that older customers wanted a luxury sport utility vehicle, features from the luxury car line could be combined with a truck chassis and an entirely new product family created.

Prior to the advent of computer models, expensive prototype vehicles would be produced and evaluated by the target customers. These prototypes were evaluated for appearance, construction, compliance with government regulations, performance, customer acceptance, and a host of other criteria. Recommended changes and improvements were identified, incorporated into the design, and additional prototype vehicles produced. This cycle was repeated until an acceptable product design was agreed upon. Once a product design concept was solidified, manufacturing feasibility and production costs could be determined.

The first step in determining manufacturing feasibility was to identify the manufacturing processes required to produce the desired product. Raw materials had to be transformed, by casting, molding or machining, some items required decorating, and components and subassemblies needed to be assembled into the final product. Knowledge of the capabilities of each of these processes was applied to decide if it was possible to produce the desired product using existing technology. If manufacturing technology did not exist, a study was performed to determine if the technology could be developed in time to support the desired introduction of the product.

Assuming that the manufacturing technology existed, the next step was to identify the rate at which the product had to be produced and the acceptable quality criteria. Again, knowledge of the manufacturing processes allowed one to predict the facilities and tooling necessary, as well as the associated human resource requirements. From all of this information, estimations of the feasibility and costs to produce a given product, at a given volume, could be made.

Take note that each phase of this development required either a physical representation of the product or process, or extensive knowledge and experience, or both. Physical mockups or prototypes were both expensive and time consuming. The changes that took place in the 1990s provided opportunities to reduce or eliminate costly prototypes, shorten the testing and approval cycles, and capture the knowledge and experience necessary to make sound judgments regarding feasibility. One of the computer tools, identified as a critical enabler to these new design methods, was the digital human model. The remainder of this chapter will explore the expectations and realities of this technology, and present a case study describing the actual use of a digital human model in product and process design.

8.2 Digital Human Modeling

Often, perceptions do not match reality. Such is the case with digital human modeling. The proliferation of video games and animation has led many to believe that accurate simulations of human motions are commonplace and simple to develop. In fact, accurate simulation of realistic human motions and, more importantly, the analysis of these motions is a complex and difficult task. While animations may provide a realistic visualization of human motions, many are captured directly from humans moving through a sequence. For the purposes of product and process design simulations, it will be necessary to identify starting and ending points in space and have the simulation generate realistic sequencing and movement paths.

In general, the technology available today generates motions that are closer to those of a robotic device. For example, when a robot moves from one point in space to another, all of the degrees of freedom are coordinated to start and stop together. This results in a smooth motion and a continuous path for the tool center point of the end effecter, which would equate to the human hand. Humans, however, do not move in this fashion. When moving the hand from a starting location to an end point, often the head moves first to identify the target position, then the hand and arm move toward the target, and finally the torso moves to increase the reach of the arm.

This is not to say that animations do not provide valuable information. Much can be detected simply through the visualization of a human form moving through the postures necessary to produce or assemble a product. However, if any analysis is to be performed by the simulation, a much more sophisticated modeling capability is required. The analysis methods for many tasks exist today and have demonstrated their value as stand-alone evaluation tools. But the most significant opportunities exist by embedding these analytical tools into the simulation environment. Engineers and process designers have indicated repeatedly that if they are to be expected to analyze for additional aspects of the process, the analysis tools cannot require them to leave their design environment to perform an analysis and then import data or move back and forth from design to analysis. The simulation environment must enable them to design work environments and analyze them simultaneously. Much has been published that could lead one to believe that this capability is available and in use today, but once again, perception does not match reality.

Another compelling concern, associated with the introduction of simulation as a primary design tool, is managing change. It is human nature to resist change and the migration to a new way of designing and testing products and processes has certainly not been spared this resistance. It has been necessary to design simulation environments that are as similar as possible to the traditional methods, while adding significant capabilities and reducing costs. Three dimensional renderings that can be easily manipulated provide the familiarity of walking around a physical model without the costs of building the model, the floor space to display it, or the security to protect it yield significant opportunities, but also represent significant change to designers and reviewers. The simulations and their development environments must reflect the major steps of the traditional design methods. That is, they must allow for review of initial concepts, intermediate reviews of overall products or processes as well as individual details, and finally, detailed reviews of completed designs. It is also important to consider the environments in which simulations are presented for review. Traditional design reviews take place in conference rooms or design studios. It is important to consider the contribution of the presentation setting to the perceived importance or sophistication of a simulation under review. Reviews should not be conducted on a desktop monitor, but rather, projected onto life sized or larger screens. And finally, computer operators must have sufficient skills and experience to manipulate displays smoothly and flawlessly.

8.3 Case Study—DEW98 Door Latch Module

The following case study presents one of the initial applications, within Ford Motor Company, of human modeling and simulation for resolution of product and process design concerns related to an automobile door assembly. This case study was used to determine the feasibility of using human modeling and simulation to conduct ergonomics evaluations.

The Vehicle Operations component of Ford has a core engineering department, staffed with a team of ergonomists, responsible for evaluating new product and process designs on new model programs. The team was requested by the door product designers to analyze the assembly of a lock module into the door structures. This assembly operation would be performed by a workforce of both men and women, ranging from very small to very large in stature. The inner door structure was a carryover design that contained access openings to enable assembly of interior components. The design group wanted to confirm that the openings were sufficient to provide arm clearance and allow smaller workers to reach areas requiring assembly operations. If the capability of the workforce was determined to be below acceptable limits, redesign of the door inner panel would be necessary, resulting in extensive additional testing for structural performance. Since concerns were raised late in the program, prototype parts were not available, and a rapid evaluation was needed.

The Vehicle Operations ergonomics team contacted the virtual factory simulation group and requested that a simulation be developed using the Technomatix, RobCad Man product. Since computer aided design (CAD) data already existed for the product, it was a relatively simple task to import the product information into the RobCad environment (Fig. 8-1). Once imported, the RobCad Man human model was introduced.

The sequence of assembly and the path that components must follow was identified in the door assembly process description. This information was then translated into motion commands for the human model. This process required approximately four days to complete, and provided the engineer with a visualization model of the expected process.

The first question to be answered using the simulation model was if the lock mechanism could be manipulated into position through the opening in the inner door panel. The clearances between the lock module and the door panels were small, with minimal clearance for the operators' hands. Since a range of operators would be performing the task, it was desirable to analyze for the largest hand size. The hand size of the 95th percentile male was selected for this analysis. Using a model with the 95th percentile male anthropometry, it was determined that sufficient clearance was provided (Fig. 8-2).

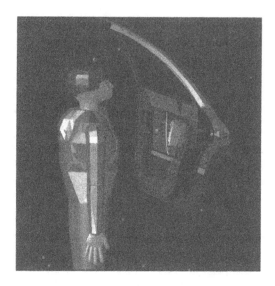

Fig. 8-1 Digital human model with CAD door model.

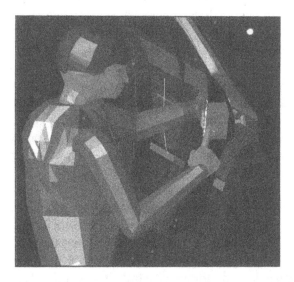

Fig. 8-2 Lock assembly to door structure.

The next question was whether the 5th percentile female would be able to reach the locations necessary to position the lock module into the final assembled position inside the door. At this point it was necessary to know the height of the door relative to the operator, in order to determine the postures required to complete the assembly. It was assumed that the lock module would be assembled to the door with the door removed from the vehicle. This process provided the opportunity to adjust the door height during the assembly process. Given this assumption, simulations were analyzed for feasibility of placement, as well as hand starting of two mounting screws to secure the lock module to the door. It was determined that the 5th percentile female could position the lock module and hold it in place while hand starting two attachment screws (Fig. 8-3).

Fig. 8-3 Reach analysis for 5th percentile female.

The conclusions of these simulations were presented to the product design engineers. At this point however, the assembly process was modified to assemble the doors while mounted onto the vehicle. As a result of this change, it was necessary to conduct an additional simulation to confirm that all operations were still feasible.

Modifications to the simulation were made to position the door onto the vehicle and position the vehicle at the appropriate height to reflect the vehicle line. Again, the 5th percentile female anthropometry was used to determine reach capabilities. At this point it was determined that the raised position of the door prevented the smaller operator from reaching the final assembly location, and hand starting of the two mounting screws was beyond the reach of smaller workers.

As a result of the simulations, it was concluded that the proposed door design was feasible for a worker population made up of a range from the 5th percentile female to the 95th percentile male, if doors were assembled in an off-line operation that allowed the door height to be adjusted. If doors were to be assembled while attached to the vehicle, the access openings in the inner door panel would have to be enlarged to accommodate the smaller workers.

The decision was made to assemble doors off-line and provide for height adjustments. The cost of this option was much less than modification to the stamping dies used to form the inner door panels and access openings.

Numerous lessons were learned from this application of human modeling and simulation. The most important lesson was the recognition of the benefit of a simulation using human models. Without this simulation, numerous physical mockups of the assembly operation would have been necessary to answer clearance and reach concerns. Multiple process options, such as doors on vs. doors off, might not have been evaluated.

8.4 Technology Assessment

The available capabilities of the human model used for this simulation were limited to visualization of postures and motions, and analysis of clearance and reach concerns. Although these limitations did not allow any biomechanical or metabolic evaluations, the insights gained provided valuable information to the design processes for both the product and the manufacturing process. It was recognized that, even though the analysis capabilities were severely limited, the benefits realized justified the costs and time spent to develop the simulation.

As a result of this pilot application, a decision was made to compare various human simulation models and select a single model for future applications. It was agreed that human modeling provides significant opportunities to shorten and improve both the product and design processes. A team of ergonomists developed a list of required features and capabilities for human simulation models and engineers and selected models were evaluated and compared. A single model was selected and a second, more in-depth, evaluation was performed. This exercise identified the strengths and areas for potential improvement of the selected modeling tool.

A vision of success for the human simulation model was developed requiring the model to possess a broad list of functional and operational characteristics. For example, the model must be able to display a realistic looking representation of all anthropometric groups from the 5th percentile female through the 95th percentile male. The model must be able to accurately calculate moment forces for all the functional joints. The model must be able to calculate compressive forces at the L4/L5 and the L5/S1 vertebra. Multiple human models must be capable of being displayed and analyzed simultaneously. Motions must be representative of realistic human motions. These characteristics, and a desire to have the model function within currently used simulation environments, define criteria for a successful human simulation modeling tool. It should be noted that none of the available human simulation models satisfied all of the wants and needs of the Ford criteria, however the "Jack"™ model from EAI provided the closest match to the vision of success, and was selected for integration and future development.

8.5 Future Direction

As the engineering culture evolves within manufacturing enterprises, acceptance of design tools such as human simulation modeling and the realization of the tremendous benefits of these design tools become more and more prevalent. Ford will continue to implement human simulation modeling and will continue to contribute to the support of the research and development necessary to improve the performance and capabilities of these technologies. Competitive pressures to produce better products, faster and at a lower cost, necessitate the embracing of all technologies that have potential to streamline the design and manufacturing processes. Technological advances in computer hardware continue to bring the cost of using advanced analytical tools and methods down, and enable a wider audience to access these concepts. And above all, the application of human simulation models has the potential to reduce work-related injuries and illnesses by reducing exposure to ergonomic stressors. This is the most significant benefit to Ford Motor Company.

Chapter 9

Summary

Don B. Chaffin, PhD
The University of Michigan
Ann Arbor, Michigan

9.1 What Did We Hope to Learn from These Cases?

As any new technology begins to be applied by various groups or individuals, it is always good to pause periodically and reflect on the collective experiences of the users of the technology. In this book, we have had the opportunity to look at seven different applications of digital human models, presented and discussed by very articulate and experienced individuals. Each of these case studies has identified several benefits and limitations of digital human modeling, which I will summarize here. However, before doing so, I'd like to briefly restate what we hoped to learn from these case studies.

The first thing we all wanted to know was whether the users truly gained some insight about a particular design that would have been technically difficult or very expensive to obtain without the use of a digital human simulation. In essence, did this approach add real value to the resulting design and engineering of some system relative to more traditional methods?

At a more detailed level, we wanted to learn which specific factors seemed most important in the application of this technology, since surveys of past users had listed many different desirable features to have in any future digital human modeling and analysis system. Some of the desirable features that are often mentioned are listed in Table 9-1.

Table 9-1 Some of the most desirable features in digital human models based on 40 responses to a survey of users performed by C. Nelson for the SAE G-13 Committee in 1996

- Selection of several different population anthropometric databases
- Inclusion of different clothing and personal protective equipment
- Prediction of population strength and endurance in manual tasks
- Accurate representation of normal human motions in dynamic tasks
- Prediction of line of sight and projecting mirror view capabilities
- Prediction of normal task performance times
- Assessments of maximum reach and obstacle interference
- Seamless integration with other CAD systems and databases

Another issue that is often raised by those considering the use of this technology is the need for the computer generated hominoid to "look and behave" like a real person. There is no doubt that this is a primary feature of any human simulation system being used for entertainment purposes. However, the general objective of the applications discussed in this book is not to entertain people with enhanced human forms and motions, but rather to improve the design of various products and processes so that these accommodate a large variety of customers and workers who must interact with these designed systems on a daily basis. Certainly it would be wonderful to have digital human models which look and behave as real people in such simulations, but perhaps it is even more important to a designer to quickly and easily perform various biomechanical, physiological, and psychological analyses of a user group. In other words, can the latter "ergonomic analyses" be obtained and used effectively without having computer generated hominoids look and behave as real people?

So, as you read the following summary of the cases, you may want to keep in mind these three general questions:

1. Did these cases provide evidence that existing digital human models can provide real value in the design of a proposed product or process that would be difficult to obtain without this technology?

2. Which particular general features of digital human models were helpful to have in these cases, and which ones were difficult to achieve or problematic?

3. Was the non-human look and behavior of the hominoid figures used in the chosen models a problem in achieving better ergonomic features for the resulting product or process designs?

9.1.1 What General Benefits Were Demonstrated in These Case Studies by the Application of Digital Human Modeling?

Our seven reference cases can be characterized as applying different types of digital human models to improve the human-hardware interfaces required in:

Chapter 2	International Space Station Design
Chapter 3	Aircraft Munitions Handling
Chapter 4	Large Ship Bridge/Helm Layout
Chapter 5	Heavy Vehicle Occupant Packaging
Chapter 6	Small Car Pedal Location
Chapter 7	Sheet-Metal Parts Handling
Chapter 8	Car Door Assembly Analysis

Among these cases, various design problems were analyzed with the assistance of a digital human model. In one case the location of a handle to steady a person from falling was a critical design issue. In another case the concern was to be able to reach an object that needed to be manipulated. Table 9-2 summarizes many of the major design issues that the authors of each case believed were successfully addressed through the use of a digital human model.

This table seems to indicate that the most prevalent use today of digital human modeling is to simulate people of extreme sizes (i.e., to perform 3-dimensional anthropometric analyses) for the purpose of providing designs wherein a large variety of people can reach, see, and/or manipulate objects. In a few cases, there existed a need to use a digital human model (DHM) for predicting a populations' reach and clearance capability, included the mitigating effects of different clothing or personal protective equipment, such as heavy gloves or helmets. In some other cases, the issue was one of how much human strength and/or endurance was required to perform a manual exertion, with special concern that the final design comply with federal NIOSH or DOT policies. In a few cases, the authors believed one of the most important features of a

Table 9-2 Summary of the design issues successfully addressed through the use of a digital human model

Design Issues That DHMs Addressed Successfully	Ch 2 International Space Station Design	Ch 3 Aircraft Munitions Handling	Ch 4 Large Ship Bridge/Helm Layout	Ch 5 Heavy Vehicle Occupant Packaging	Ch 6 Small Car Pedal Locating	Ch 7 Sheet Metal Parts Handling	Ch 8 Car Door Assembly Analysis
1. Visualize extreme populations (small F or large M) for reach/visualization	√	√	√	√	√	√	√
2. Hand clearance for maintenance/operation	√	√		√			√
3. Grab handle/step locations	√	√		√			
4. Extreme environment (rain/snow)	√		√	√			
5. 2+ person coordination communication for large mockup analysis			√			√	
6. Seated vs. standing operation			√				
7. Seated pedal locations - leg postures				√	√		
8. Population strength - compare policies of NIOSH, DOT compliance policies		√		√		√	√
9. Visualize - communicate ergonomic issues		√	√			√	√
10. More creative designs - 'What Ifs'				√			√
11. Combined product and manufacturing concurrent engineering							√

DHM was that the simulations and associated graphics allowed both product and process designers to understand better the potential problems a particular population subgroup could have when operating or servicing a proposed design. This last point was emphasized by some of the authors in that they provided estimates showing that the use of a digital human model saved many months and thousands of dollars in design and prototype testing, compared to their traditional methods.

9.1.2 What General Limitations Were Found in the Application of Existing Digital Human Models?

Despite the many positive outcomes, all of the cases illustrated certain limitations of the existing technology—limitations that the authors believed must be overcome to allow a much wider acceptance and use of the technology in the near future. Table 9-3 summarizes some of these limitations.

Table 9-3 Summary of major limitations of digital human modeling as discussed in each case study

DHM Limitations and Concerns	Ch 2 International Space Station Design	Ch 3 Aircraft Munitions Handling	Ch 4 Large Ship Bridge/Helm Layout	Ch 5 Heavy Vehicle Occupant Packaging	Oh 6 Small Car Pedal Locating	Ch 7 Sheet Metal Parts Handling	Ch 8 Car Door Assembly Analysis
1. Creating 3D DHM within a CAD environment	√	√		√	√		
2. Modeling exertions in extreme postures	√			√		√	
3. Deriving postures or motions for dynamic analyses from motion capture files	√	√		√	√		√
4. Providing control-forces needed to perform dynamic analyses			√	√	√	√	√
5. Estimating postural comfort - for long periods			√		√		
6. Selecting "critical" tasks				√			
8. Having large screen 3D projections for group discussions - design reviews							√

One of the most frequently discussed limitations in the use of digital human modeling is in the difficulty with obtaining the necessary input data for a complete analysis, or with imbedding the digital human model into an existing CAD model, which could then provide much of the needed input data. The problems with data access and software compatibility are of course very common in large scale simulation and CAD systems. Such problems are perhaps exacerbated in these types of cases, however, by the unique and complex nature of the data necessary to perform a complete human simulation for ergonomic assessments. For instance, not only must the designer specify where a person might stand or sit, but also where objects are located that could be reached, seen, and manipulated by a given population of interest. In other words, to simply begin to specify the geometric input data necessary to use a digital human model, the designer must be able to envision how a task might be performed and what conditions might enhance or detract from such performance. And the geometric data describing a vehicle interior or work environment is only part of the data necessary for a complete ergonomics analysis. Because most people operate well within a rather limited physiological and psychological spectrum of environmental stressors, the designer must also be able to supply data about the repetition or length of time a particular task is to be performed, or what hand force must be applied to move an object, or whether the floor or handle is slippery, or if the temperature or lighting is sufficient, or whether small objects or letters can be distinguished from their background. Once these types of data are acquired they must be provided to the model in a way that allows the designer to run alternative "what if" analyses. In other words, future digital human models must allow the user to quickly and easily access a great deal of geometric and human performance data.

Positioning the hominoid correctly was also a major issue. Many of the case study authors believed that having valid posture and motion prediction capability would greatly improve the ease of use of their particular digital human models. They seemed concerned that many future users of this technology won't have the training or experience necessary to be able to accurately move and position the hominoids correctly within a particular physical environment being studied; and pointed out that this limitation is most important in those situations when a proposed design must accommodate a large variety of people, those that are large or small, men or women, and young or old. In other words, they seemed to express the opinion that the average designers of a new system could hardly be expected to know how their proposed designs could accommodate the motions and postures of an "average" person, not to mention extreme populations. They believed that such knowledge must be provided to a great extent by the particular digital human model being used.

In this latter context, they acknowledge that some inverse kinematics algorithms are now being developed and imbedded in the DHMs to assist the designer in manually positioning and moving the hominoids, but unfortunately the resulting postures and motions are still very robotic and "inhuman" in form, and/or require considerable computer run time to generate. This has limited the use of the DHM technology to analyzing a sequence of single tasks or exertions that often are static in nature, rather than simulating a series of dynamic movements for an entire job or activity of interest.

It should be clear from the case studies presented that to allow digital human modeling to move beyond its current single task analysis orientation will require several major enhancements to the technology. Specifically, we will need to have: (1) a very large database and models of many different and common human motions; (2) algorithms for easy and accurate modification and blending of the existing motion data; and (3) a common input language, or dialog standard to allow users to easily specify and string together motions into a complete activity or job simulation. These are not new requirements, and were well discussed in the text by Badler, Phillips, and Webber in 1993 [1]. The ever increasing computational speeds of computers today certainly would indicate that algorithms to provide such a desirable feature should be computationally feasible in the very near future, provided researchers continue to develop motion databases and dynamic human motion models for this purpose, as discussed in Chaffin, 1999 [2].

Another generally desirable but not yet implemented feature of the present digital human models is their use in collaborative design. The case involving the automobile door assembly (Chapter 8) illustrates the need for several different groups of people (product designers, manufacturing engineers, equipment vendors, and production personnel) to be simultaneously aware of the impact of their ideas on those people who are destined to use and operate the systems they specify and design. It appears that if various human-hardware interface issues of a proposed design could be simulated in real time and the results shared with all of the appropriate people on-line, perhaps even in a 3-dimensional, full scale, virtual projection, then concurrent design processes should result in better ergonomics designs at an early phase of the design cycle. At a minimum, simultaneous group presentations of digital human simulations will assure that in the future the typical design reviews conducted today will include human-hardware ergonomics assessments which are too often performed near the end of the product design cycle.

9.2 Some Concluding Thoughts About DHM Applications

It should be clear from these seven case studies that the present technology of digital human modeling is addressing many different types of important human-hardware issues during the design of new products and manufacturing processes. These cases also demonstrate, however, that to use the technology successfully one must have a great deal of data about the geometric environment in which the DHM is imbedded, as well as other data about the specific tasks and environment in which a person must operate. In essence, the use of a DHM requires the designer to be very specific about all types of attributes of a proposed design that could affect a person when operating or servicing a proposed system. In this context, one must conclude that the technology is not very forgiving at this time. It requires the user to be quite knowledgeable about many different human factors and ergonomic principles to avoid misusing a DHM in a proposed design.

So in returning to the three general questions posed earlier in this chapter, it appears that the answer to the first question about whether the DHM users found real value in the use of the technology is affirmative. They all managed to solve important questions with the technology, and a few of them even stated that this approach saved significant time and money compared to their more traditional methods.

The second general question concerned which types of human-machine interface problems are being chosen for application of this digital human modeling technology. Based on these seven cases it appears that we are most often using the technology to evaluate human "fit, clearance, and line of sight issues." In essence the technology is seen as a replacement for 2D and 3D physical manikins to better solve complex anthropometric problems. These new computerized manikins can now be much more elegantly scaled to represent the shape and size of a very diverse population, and can be more easily manipulated within an environment of interest than existing physical manikins placed in prototypes of a proposed design.

The next most popular use of the technology appears to be in solving problems related to the strength of various people performing manual exertions, and assessing comfort or endurance. In general, these are much more difficult ergonomic attributes to analyze than to assess the effect of a person's size and shape on a proposed design. This complexity may be the reason why such attributes are not yet often considered in DHM applications, but this should rapidly change as designers gain more confidence in modeling such human attributes. The same is true of modeling human motions. Many of the case authors acknowledged the need for this feature in their simulations, but because it is not readily available, or the resulting motions were not very humanlike, they did not use it.

This last point leads to the final question about the need in DHM analyses to have hominoids that look and behave like real people. Several of the case authors wanted normal motion and posture generation capability to assist them in their use of a DHM for solving a particular ergonomics problem. Yet they apparently were able to approximate the necessary postures to solve the problem of interest without this feature. In essence, they appeared to be able to use their own training and experience in human kinesiology to overcome the motion prediction deficiencies in their DHMs. So the answer to the general question of look and motion behavior would appear to be that it is highly desired to enhance these features in future models so that the models are easier to use, especially by non-ergonomics trained people, but if the person using the model is skilled in postural synthesis, then these features are not necessary to perform an effective ergonomics analysis of a task. The need for the hominoid to look very lifelike did not seem to be a major issue. However, for general presentations of one's work with a DHM, it might be very helpful to have the hominoid look like a real person, i.e., it would help in the communication of one's results to others not as familiar with the technology.

So what did we learn from these cases? First, the technology of digital human modeling is moving into mainstream design of vehicles and workplaces. Second, digital human modeling is able to shorten design-to-build process times and costs. Third, the initial applications of the technology are mostly sophisticated anthropometric analyses to assure that both large and small individuals are well accommodated. And fourth, much more knowledge about human behavior and biomechanics needs to be added to the present DHMs to allow them to be more easily used and of greater value in computer aided design systems. In this latter regard, it would seem that the DHM users did not require the technology at this time to decide on questions about perceptual or cognitive capabilities of people. This may due to the inability of the present models to predict such capabilities, though some very good perceptual acuity models exist in the human factors literature. Or it may be simply a matter of priorities, in that issues relating to the physical accommodation of people are so important now that the DHM users did not have the time to deal with other types of human performance issues. Undoubtedly this will change in the future. Questions of whether various people can see, hear, and otherwise identify various attributes of a virtual environment, or for that matter decide on how to react to certain changes in a virtual environment will become important, and sooner than we may wish to acknowledge. A simple example of this trend is happening in the design of automobile interiors, where-in many different and new electronic devices are being quickly added to "assist" drivers and occupants. These devices not only need to be placed so people can reach them, but each device must be designed to allow the user to easily understand and use various I/O functions with a minimum of training, and perhaps while speeding down a crowded highway. When one thinks about these types of ergonomic issues it should become painfully obvious that the present DHM technology has just begun to develop into a useful expert system for ergonomics practitioners and system designers. Though the cases in this book have demonstrated a very important role for DHM simulations in design, it should be clear that we have only been given a small view of the future potential for this technology. A future that undoubtedly will be very exciting for all those concerned with designs that better serve people in all walks of life.

References

1. Badler, N.I.; Phillips, C.B.; and Webber, B.L., *Simulating Humans: Computer Graphics Animation and Control*, Oxford University Press, New York, 1993.
2. Chaffin, D.B.; Faraway, J.; and Zhang X., "Simulating Reach Motions," SAE Human Modeling for Design and Engineering Conference, The Hague, The Netherlands, 1999.

Glossary

accelerator-heel point Assumed point of contact of driver's heel with floor when foot is placed on the undepressed accelerator. [1]

animation-based simulation Simulation where the most important output is the animation of real world activities as is often the case with digital human simulations. Other types of simulation may use animations to augment or clarify their outputs.

anthropometric database A database that contains measurements of the height, weight, build, and other physical dimensions of a group of people.

anthropometric variability Natural variability in the physical dimensions of a group of people.

anthropometry Study of the measurement of the human body. *See also* ergonomics. [1]

anthropomorphic model A physical or electronic representation of characteristics (anthropometric, biomechanical, cognitive, etc.) of a human being (or population) that is used for simulation, testing, or evaluation in analysis or design processes.

articulation A joint, as the joining or juncture of bones of the movable segments of the human. [3]

avatar A virtual representation of a real person and associated motions. Also called hominoid.

biomechanical model An analytical or computerized representation using laws of physics to describe the moments and forces resulting from body motions and exertions.

biomechanics The study of the effects of internal and external moments and forces on the human body in movement and during exertions. [7]

Boeman Computer simulation that grew out of the "First Man" program, perhaps the first attempt to develop a computer simulation of a person performing a reach task.

B-pillar Vertical cab member, located rearward and outward of the driver and passenger's positions, that provides structural support and strength for the door latch, occupant restraints, and roof.

clearance envelope With regard to a tool, the volume around the fastener head that should be free in order to provide crewmembers with optimal physical clearance for tool actuation.

collision prediction The ability in a simulation system to determine when two three-dimensional objects will intersect.

computer aided design (CAD) Describes any or all of the multitude of graphic, computational, and engineering design functions performed with the aid of a computer and leading to the choice of a final design. These range from the aesthetic to the most detailed analytical computations. [2]

computer aided engineering (CAE) The use of computer-based tools to assist in solution of engineering problems. [4]

computer simulation A logical-mathematical representation of a simulation concept, system, or operation programmed for solution on an analog digital computer. [5]

computer-assisted ergonomic analysis A type of ergonomic analysis in which computer technology, rather than traditional methods such as wooden mockups, is used to verify the ergonomics of a design.

concurrent engineering The simultaneous design and development of a product and the process to produce it in adequate numbers. Also called simultaneous engineering or integrated engineering. [2]

Crew Chief A 3D computer-aided design (CAD) manikin developed during the early 1980s by AMRL; successor of COMBIMAN.

cycle time The total time required to complete all of the individual tasks for a specific operation. The time from the beginning of an operation until the beginning of the next repetition of that operation.

decision threshold A "flag" applied to an anthropomorphic model, for example, a percent of the population capable or a cutoff value based on some consensus of stress level.

digital human modeling Computer simulation of human motions and exertions.

digital manikin Static representation of a human being derived from a digital human model.

digital mockup (DMU) A computer-generated representation of a workspace, assembled using electronic drawings from a CAD system.

egress *1.* The action or right of going or coming out. *2.* A place or means of going out.

emergency operations Emergency tasks on the International Space Station (e.g., retrieving and activating a portable fire extinguisher).

energy analysis A type of ergonomics analysis where the metabolic energy associated with loads, distances, lifting and carrying demands, and other such factors are determined.

enfleshed Describes a human simulation model that includes realistic flesh.

ergonomic analysis Analysis of the interactions between a worker and his environment.

ergonomics The study of work in relation to the environment in which it is performed and the people who are doing it using the application of scientific techniques to improve performance and safety. [2] *See also* anthropometry.

ergonomist One who specializes in ergonomics.

expert system In relation to ergonomics, a software system that is able to provide advice about how certain groups of end users of a proposed design will be affected by the design.

eye reference point In digital human modeling, the point from which the manikin is able to gain visual access to a control or component. An eye reference point can also serve as the manikin attach point within the digital mockup.

First Man program Sponsored by the U.S. Naval Air Development Center (NADC), perhaps the first attempt to develop a computer simulation of a person performing a reach task, performed for the Boeing Aircraft company in the late '60s; later became "Boeman."

Fitts' Law A method of assessing speed and accuracy of performance. Relates movement time between a starting point and a target to the distance traveled during the movement and the size of the target. [8]

force feedback device Virtual reality equipment that allows the user to feel objects handled by the human model.

functional analysis A systematic and formal procedure used by human factors professionals to identify and allocate specific functions to components of a complex system, including human tasks.

Garg Energy Model A model that predicts the overall and elemental metabolic energy requirements of a sequence of activities associated with manual materials handling.

haptic feedback In a virtual environment, feedback relating to the sense of touch.

head mounted display A device for displaying a virtual environment where computer graphics are positioned just in front of the eyes.

hip point *or **h-point*** Defined by SAE Surface Vehicle Recommended Practice J1100, Motor Vehicle Dimensions, as the pivot center of the torso and thigh on the two- or three-dimensional devices used in defining and measuring vehicle seating accommodation, per SAE J826. Society of Automotive Engineers, "SAE Surface Vehicle Recommended Practice J1100, Motor Vehicle Dimensions," Warrendale, Pa., 1999.

hominoid *See* avatar.

human factors The physical and psychological requirements of human beings that must be considered when designing products in order to make the products practical, efficient, safe, and easy to use.

human factors engineering The area of knowledge dealing with the capabilities and limitations of human performance in relation to design of machines, jobs, and other modifications of the human's physical environment. [4]

human simulation model A digital representation of the human form, capable of displaying and analyzing a variety of realistic motions, postures, and actions.

human simulation technology Technology for simulating a person functioning in a particular CAD environment, such as walking, reaching, carrying, or exerting force on an object.

hybrid mockup A digital mockup that combines a simple wooden mockup with a virtual environment that is exactly lined up. This introduces haptic feedback into the virtual environment.

inertial load Forces with which a human operator must contend due to the mass or movement of an external object.

intelligent assists Powered devices that use sensors and servo controlled motors to assist human operators with the manipulation of tools, parts, or materials through paths with computer controlled speed, motion, acceleration and deceleration, and target placement.

inverse kinematic algorithm An algorithm developed in the late '80s to assist a user in posture predictions; used in biomechanical model of population force prediction.

inverse kinematic posture prediction Posture prediction by the inverse kinematic algorithm.

ISS Node 1 International Space Station Node 1, the first Space Station node and the first major U.S.-built component of the station.

Jack Name of the human simulation model from Unigraphics, formerly EAI (Engineering Animation Inc.).

joint load moments The load moments at all the major body joints, produced by hand forces and body segment weights when a person performs an exertion.

kinematics The branch of mechanics dealing with the description of the motion of bodies or fluids without reference to the forces producing the motion. [5]

knee height sitting The height of the top of the knee measured from the floor (with or without shoes) when a person is sitting erect with feet flat on floor and the lower legs vertical.

Liberty Mutual (Snook) Tables Tables that provide maximum acceptable lift/push/pull forces for a variety of tasks.

lumbar spine The portion of the spine between the lowest ribs and the pelvic girdle.

maintenance operations Maintenance tasks on the International Space Station (e.g., remove and replace lamp housing).

manikin An electronic 3-D computer definition of a human form. A manikin can be posed and articulated to simulate movement.

motion models The task simulation capability of DEPTH, which allows fairly complex tasks to be automated using a hierarchy of subtasks.

multivariate approach An approach to digital human simulation in which manikins are developed using boundary conditions of key anthropometric dimensions, rather than the traditional univariate anthropometric percentiles (i.e., 5th, 50th, and 95th).

multivariate correlation Correlation involving a number of independent mathematical or statistical variables.

natural language commands A language spoken or written by humans, as opposed to one used to program or communicate with computers. Given the complexity, irregularity, and ambiguity of human language, natural language understanding has been a challenging goal in computer science.

neutral body posture (NBP) The relaxed, unrestrained posture the human automatically assumes in weightlessness.

NIOSH Lift Equation An equation that is used to evaluate the safety of a lifting task; takes into consideration the initial and final load locations, and the frequency and duration of lifting to predict population lifting capabilities.

non-linear optimization A procedure for locating the maximum or minimum of a function of variables that are subject to constraints, when either the function and/or the constraints are nonlinear. [5]

normal operations Operational tasks on the International Space Station (e.g., video camera operation).

physical mockup A physical representation of the work environment made of building materials such as wood.

prototype A model suitable for use in complete evaluation of form, design, and performance. [4]

psychomotor behavior Behavior representing both mental and motor activity.

psychophysical data Data collected by having subjects select their maximum acceptable load of handling under experimental conditions. Subjects adjust the load handled until they feel it is not excessive. These selected loads represent design guidelines. [7]

range of motion The degree of movement of segments that can occur at a joint. [4]

Rapid Upper Limb Analysis (RULA) A method to assess the overall stress associated with a materials handling task.

reach capability The maximum distance a human can safely extend the hand while supporting a given load.

reach envelope A geometric representation showing the reach capability for a certain manikin.

reach strategy The selected path, of various options, through which the hand is moved from one location to another.

reach trajectory The path that the hand follows when moving from one location to another.

ROBCADMAN Human simulation model by Technomatix.

simulation Using computers, electronic circuitry, models, or other imitative devices to gain knowledge about operations and interactions that take place in real physical systems. [5]

simulation technology Software, hardware, and other devices used for simulation.

sitting height A measure of the vertical distance (taken along the back) from the table surface to the crest of the head as the subject sits erectly on the table, knees pressed against the table edge and head in the eye-ear horizontal plane. [4]

somatotype A basic body type; three primary components are ectomorph, mesomorph, and endomorph. [4]

spinal lengthening The "growth" in height that occurs in a 0-G environment.

static capabilities Population maximum voluntary exertion values under static conditions.

stature A measure of the distance from the floor to the vertex of the head, taken either front or back as the subject stands erectly with heels together. [4]

stereolithography A method of making engineering models directly from computer aided designs. It is based on a multi-slice CAD model being translated, which can be reproduced in an acrylic or photopolymer material. These are then adhered together with a laser to form the required three-dimensional part. [2]

tag point A target point or coordinate system that serves as the manikin's goal or reach point.

target-specific reach path The actual trajectory of the fingers, hand, and arm while reaching for a specific target.

task-based strength algorithm A method of determining how much strength a human has in performing a specific task or job. This differs from other strength algorithms that use load predictions for relevant joints. Task-based strength algorithms are typically only valid for one specific task whereas joint-based strength algorithms can be applied to many different situations. It can be argued, however, that task-based strength algorithms are more accurate than joint-based predictions for the tasks they were developed for.

time study A work measurement technique, generally using a stopwatch or other timing device, to record the actual elapsed time for performance of a task, adjusted for any observed variance from normal effort or pace, unavoidable or machine delays, rest periods, and personal needs. [4]

Unity Node International Space Station (ISS) Node 1.

validation The process of demonstrating, through testing in the real environment, or an environment as real as possible, that a system satisfies the user's requirements. [5]

virtual environment A computer simulated environment that does not physically exist, but is experienced through sensory inputs such as sight, sound, temperature, etc.

virtual human A human model that is able to simulate movement and task performance. Also called digital human.

virtual prototyping Refers to the process of creating 3D virtual images of objects, which then can be perceived by the designer and others to assess whether the object meets certain visual design goals.

visibility plot A typically two-dimensional representation of what a human can see. These plots can outline what is and is not obscured by head gear (for example, goggles), what the human is focusing on (foveal vision), or the unobscured peripheral view.

Zarya The Russian-built control module for the International Space Station.

References

1. Don Goodsell, *Dictionary of Automotive Engineering, Second Edition*, Society of Automotive Engineers, Inc., Warrendale, Pa., and Butterworth-Heinemann, Oxford, 1995.

2. G.H.F. Nayler, *Dictionary of Mechanical Engineering*, Fourth Edition, Society of Automotive Engineers, Inc., Warrendale, Pa., and Butterworth-Heinemann, Oxford, 1996.

3. *Merriam Webster's Collegiate Dictionary, Tenth Edition*, Merriam-Webster, Inc., Springfield, Mass., 1997.

4. *McGraw-Hill Dictionary of Scientific and Technical Terms, Fourth Edition*, Sybil P. Parker, Editor in Chief, McGraw Hill Book Company, New York, 1989.

5. *SAE Dictionary of Aerospace Engineering, 2nd Edition*, Joan L. Tomsic, Editor, Society of Automotive Engineers, Inc., Warrendale, Pa., 1998.

6. *Computer User High-Tech Dictionary*, http://www.computeruser.com/resources/dictionary/index.html.

7. *Workplace Health, Safety, and Compensation Commission Ergonomics Glossary*, http://www.whscc.nf.ca/ohs/gloss.html.

8. *Miami University Human Factors/Ergonomics Program Glossary of HF/E Terms*, http://miavx1.muohio.edu/~psy4cwis/ergocenter/glossary.html.

Index

Abbreviations are used to indicate figures (*f*) and tables (*t*).

About the Author

Don B. Chaffin received his PhD in industrial engineering from the University of Michigan in 1968. He has held faculty positions at the University of Kansas, University of California at Irvine, and the University of Florida. He currently is the Lawton and Louise Johnson Professor of Industrial & Operations Engineering and Biomedical Engineering at the University of Michigan, where he directs the Human Motion Simulation Laboratory. His primary research concentrates on the biomechanical aspects of human exertions and motions to enhance computerized workspace and vehicle design methods. He has received awards from the HFES, ASB, SAE, AOEM, AIHA, and ISB, and has been elected to the National Academy of Engineering.

Don Chaffin's email address is dchaffin@umich.edu.

About the
Contributing Authors

Chapter 2, "Anthropometric Analyses of Crew Interfaces and Component Accessibility for the International Space Station"

Cynthia A. Nelson is a principal engineer/scientist in the Human Factors and Ergonomics Technology Group of The Boeing Company in Southern California. She has 18 years of aerospace experience and has authored over thirty publications. Ms. Nelson serves as chair of SAE's G-13 Human Modeling Technology Standards User Requirements Subcommittee. She holds a master of science degree in industrial psychology and is a certified professional ergonomist.

Cynthia Nelson's email address is cynthia.a.nelson@boeing.com.

Chapter 3, "Human Model Evaluations of Air Force System Designs"

John Ianni is an authority on human modeling and, in particular, maintenance simulation. He has been the chairman of the SAE Human Modeling (G-13) Software and Virtual Reality Subcommittee since 1997. Since joining the Air Force Research Laboratory in 1985, he has led several research projects with budgets totaling over $20 million. Mr. Ianni has a master's degree in computer science from Wright State University and a bachelor's degree in computer information systems from the Ohio State University.

John Ianni's email address is John.Ianni@afrl.af.mil.

Chapter 4, "Ship Bridge Design and Evaluation Using Human Modeling Systems and Virtual Environments"

Aernout Oudenhuijzen was born in Neede in the Netherlands and has worked for TNO HRFI since 1996. He began using human modeling technology while obtaining his degree in industrial engineering at the Technical University of Delft. Working for Fokker Aircraft, where he started in 1991, he was responsible for the human factors engineering of the F50, F70, and F100, and was involved in the development of new products. At TNO HFRI he worked on the design and the development of several workspaces. Among these workspaces were car interiors, ship bridges, flight decks, command and control centers, and the interior of armored vehicles. He entered the SAE G-13 Human Modeling Technology Standards Committee in 1993 and is lead for the verification and validation project for human models. He was general chairman of the DHMC 1999 held in The Hague in the Netherlands.

Aernout Oudenhuijzen's email address is oudenhuijzen@tm.tno.nl.

Chapter 5, "Using Digital Human Modeling in a Virtual Heavy Vehicle Development Environment"

Darrell S. Bowman received a master of science degree in industrial and systems engineering from the Virginia Polytechnic Institute and State University. From 1996 to 1998, Mr. Bowman worked in Patuxent River, Maryland, as a human factors engineer with the Naval Aviation Warfare Center, Aircraft Division. Currently, he is a senior human factors engineer with the International Truck and Engine Corporation in Fort Wayne, Indiana. His work at International focuses on the development and application of ergonomic analysis tools and methodologies, including human digital modeling technologies, to support the design of heavy truck and buses.

Darrell Bowman's email address is Darrell.Bowman@Navistar.com.

Chapter 6, "The Determination of the Human Factors/Occupant Packaging Requirements for Adjustable Pedal Systems"

Dr. Deborah D. Thompson is the president and founder of Plumb Ascension Consultants, LLC, an engineering and technology consulting firm specializing in process and technology integration. Previously, she served as a product development specialist with DaimlerChrysler Corporation, where her duties involved the development and integration of simulation technology within the product development and manufacturing processes and in assessing the ergonomic design of potential program vehicles. Dr. Thompson received her PhD in industrial and operations engineering from the University of Michigan.

Deborah Thompson's email address is plumb@plumb-ascension.com.

Chapter 7, "Ergonomics Analysis of Sheet-Metal Handling"

Brian Peacock obtained his PhD in ergonomics from Birmingham University. He is a registered professional engineer and certified professional ergonomist. He spent fourteen years in academia, in England, Hong Kong, Canada, and the United States. In 1986, he joined General Motors where he first worked in product design and later in manufacturing ergonomics. In 2000, he joined the National Space Biomedical Research Institute as human factors discipline coordinating scientist for NASA. He has published numerous articles and presentations on applied ergonomics, including books on statistical distributions and automotive ergonomics. He is an active contributor to the administration of the profession through service as director of the Ergonomics Division of IIE, vice-president of BCPE, and as a member of the executive council of HFES.

Brian Peacock's email address is brianpeaco@hotmail.com.

Chapter 8, "Digital Human Modeling for Improved Product and Process Feasibility Studies"

Glenn Jimmerson currently works in the Advanced Manufacturing Engineering Division of the Ford Motor Company and is located at the Advanced Manufacturing Technology Development Center in Redford. Until May of this year, when he became a Six Sigma Blackbelt, he was the section supervisor in the Advanced Assembly Technologies Department, responsible for the development and implementation of advanced assembly methods and strategies. Since 1990, he has led the research and development efforts to support the integration of ergonomics into analysis tools and engineering methodologies for the Ford ergonomics process, worldwide. For the past two years he has been the chairperson of the Industrial Advisory Board to the University of Michigan, Human Simulation Laboratory. Mr. Jimmerson has a bachelor of applied science degree in industrial electricity and has extensive experience in robotics, machine vision, and the application of artificial intelligence to manufacturing. He has more than thirty years of experience with Ford Motor Company, ranging from UAW skilled trades through his present position in advanced engineering development.

Glenn Jimmerson's email address is GJimmers@Ford.com.